VGM Opportunities Series

OPPORTUNITIES IN
TELECOMMUNICATIONS CAREERS

Jan Bone

 **VGM Career Horizons**
a division of *NTC Publishing Group*
Lincolnwood, Illinois USA

ABE3171

FEB 1 9 1996

Cover Photo Credits:
Upper left photograph courtesy of Sauder Woodworking company. All other
photographs courtesy of DeVry Inc.

Library of Congress Cataloging-in-Publication Data
Bone, Jan.
    Opportunities in telecommunications careers / Jan Bone
      p. cm. — (VGM opportunities series)
    Includes bibliographical references.
    ISBN 0-8442-4587-9 (hard) — ISBN 0-8442-4588-7 (soft)
    1. Telecommunication—Vocational guidance. I. Title.
II. Series.
TK5102.6.B66   1996
384'.023—dc20   95-24092
                                            CIP

Published by VGM Career Horizons, a division of NTC Publishing Group
4255 West Touhy Avenue
Lincolnwood (Chicago), Illinois 60646-1975, U.S.A.
© 1996 by NTC Publishing Group. All rights reserved.
No part of this book may be reproduced, stored in a retrieval
system, or transmitted in any form or by any means,
electronic, mechanical, photocopying, recording or otherwise,
without the prior permission of NTC Publishing Group.
Manufactured in the United States of America.

5 6 7 8 9 VP 9 8 7 6 5 4 3 2 1

CONTENTS

ABOUT THE AUTHOR

Jan Bone believes an interest in telecommunications pays off.

When General Motors hired Jan Bone by fax to write a story on environmental issues for one of its magazines for retired employees . . . when Motorola asked Jan to submit a resume by electronic mail for consideration for a writing project . . . when Jan and her husband were escorted around Kuala Lumpur, Malaysia, by an English teacher she'd "met" through an Internet listserv . . .

Let's just say that Jan, who spent a college-vacation summer running a PBX board for a six-floor Pennsylvania department store, manually connecting each call with a plug cord, appreciates today's technology. She marvels at how quickly technology has advanced, especially since she was 10 years old before she saw her first dial telephone.

In the small Pennsylvania coal town in which Jan Bone lived before World War II, all calls went through a central switchboard and were manually connected by operators. In fact, Jan's father had worked his way through college by serving as the sole night switchboard operator for an entire town, sleeping at the telephone company office and waking up to handle each call as it came through.

Today, Jan and her husband are a two-computer family, with local area network (LAN) software and cables connecting both personal computers so they can exchange data. Electronic mail (e-mail) allows her to send facsimile transmissions from her home computer without having a fax board installed in the computer. E-mail has also let her "interview" several experts for this book at their convenience—and hers—without having to coordinate schedules for phone conversations.

As a freelance writer and educator, she is on-line with writers, editors, and teachers worldwide. Currently she is using telecommunications technology to create a series of language lessons for Japanese students learning English, sending material over the Internet to an editor in Kyoto. Since 1987, she has been Senior Writer on the Chicago *Tribune*'s Special Advertising Sections, writing short stories about advertised products and services and sending them via modem directly into the Trib's mainframe computer.

A prolific freelance writer, she has written for such diverse publications as the *National Enquirer, Bank Administration, Woman's World,* and *Family Circle.* She has written for *Safety and Health, Today's Supervisor, Today's Worker,* and *Product Safety Up-to-Date,* all published by the National Safety Council—an organization for which she also wrote the chapter on automated manufacturing in its *Accident Prevention Manual,* tenth edition. She has also photographed and written about food manufacturing plants in Russia, Singapore, and Hong Kong.

Jan has been writing professionally since 1947, ever since her first newspaper job on the Williamsport (Pennsylvania) *Sun.* That summer's experience taught her that writing was fun, profitable, and an endless source of new ideas and information.

Between 1977 and 1985, Jan was an elected member of the board of trustees of William Rainey Harper College in Palatine, Illinois, and served as its secretary from 1979–85.

She is co-authoring (with Ron Johnson) *Understanding the Film, A Beginner's Guide to Film Appreciation* (NTC, fifth edition, 1996). In the VGM series, she has written second editions of *Opportunities in Film Careers, Opportunities in Cable Television, Opportunities in Computer-Aided Design and Computer-Aided Manufacturing (CAD-CAM),* and *Opportunities in Robotics Careers* as well as *Opportunities in Laser Technology Careers* and *Opportunities in Plastics Technology Careers.*

Since 1983, Jan has been listed in *Who's Who of American Women.* She has won local, state, and national writing awards. She is an associate member of the Society of Manufacturing Engineers and is active in the Independent Writers of Chicago (IWOC) organization.

A graduate of Cornell University with an M.B.A. degree from Roosevelt University, Jan has taught writing to adults for over 20 years. She is an instructor at Roosevelt University, membership chair for the International Consortium of the National Council of Teachers of English, and publicity chair for the Independent Writers of Chicago.

She is married, mother of four married sons, and grandmother of Emily, Jennifer, and Jeffrey—who send drawings from Florida and California by fax to Jan's Illinois home.

And she welcomes letters from the readers of this book addressed to **janbone@delphi.com,** her Internet location.

ACKNOWLEDGMENTS

The author acknowledges with thanks the help given by the following persons and organizations:

Christine Walsh Angelos, Mitzi Babka, Sally Baylaender, Dan Behm, Dan Brinegar, Thomas Douet, Leigh-Anne Gauvin, Gracemarie Howland, Charlene Johnson, Joyce A. Johnson, Shirley Lanier, Dawn-Marie Look, John Parmenter, Karol Kline Petit, Jim Reiss, Stan Simbal, Esther Sutrick, Dean Tarrell, Karen Varnas, and Shelly Weide. Special thanks to Gary S. Hutchins.

Also: American Facsimile Association, American Physical Society, American Telemarketing Association, AT&T, AT&T Bell Laboratories, Australian Embassy, Australian Trade Commission, British Telecom, Cellular Telecommunications Industry Association, Committee on the Status of Women in Physics of the American Physical Society, Delphi Internet Services, Direct Marketing Association, Electronic Industries Association, Electronic Mail Association, Government of Canada, Hewlett-Packard, International Communications Association, International Teleconferencing Association, MCI Communications Corp., Motorola Inc., National Association of Telecommunications Officers and Advisors, National Cable Television Association, National Telecommunications & Information Administration, North American Telecommunications Association, Pro Systems, Inc., Society of Women Engineers, United States Telephone Association.

FOREWORD

Welcome to the expanding and exciting world of telecommunications careers! Unsurpassed in the variety of professional options and opportunities it offers, telecommunications is growing exponentially—and that growth will continue well into the twenty-first century. In 1993 alone, 28 million new phone lines were installed globally; an additional one billion new accounts are anticipated by the year 2000. Advances in telecommunications have affected business, entertainment, and communication around the world—and new products, processes, and applications are just around the corner.

Completely revised and updated, this new edition of *Opportunities in Telecommunications Careers* discusses all of the major professional options in this rewarding field. From cutting-edge careers in research and development to behind-the-scenes support professions, this book will show you how to get started, where to find the training and preparation you'll need, and what you can expect on the job. It will serve as an essential roadmap as you find your niche and find your way in telecommunications.

The Editors of VGM Career Horizons

CHAPTER 1

TELECOMMUNICATIONS: THE TECHNOLOGY OF TOMORROW

The Information Revolution is changing lives forever. Wave-of-the-future technology is blending the written word in computerized form with data communications, video, audio, and photography in digital interactive media. At the heart of these merging technologies lies telecommunications—the thread that links them together.

Certainly Alexander Graham Bell, who spilled battery acid on his trousers and cried, "Mr. Watson! Come here; I want you!" into his primitive transmitter in 1876 could never have envisioned today's already-global environment and its implications for the future. Financial institutions authorize the movement of hundreds of millions of dollars daily. Companies like Dell invest $50 million in information technology each year, linking computers through its phone systems and getting 35,000 calls and e-mail messages every day. When voice messages, faxes, and e-mails are combined, Sun Microsystems has 1.5 million internal messages daily—120 per employee—and up 50 percent in an 18-month period.

American Express, which spends $1 billion a year on information technology, plans to put a thousand or more telephone order-entry employees on-line from home, using the electronic highway to boost productivity. Estimated additional revenue for each telecommuting employee: $30,000 each year, plus $4,500 in unused leased office space.

For less than $50 per month, Individual, a customized news source, will fax or e-mail subscribers daily with custom-sorted news, electronically searched from 300 sources. Store-forward fax, audio messaging, wireless mobile communications—all these help people connect with others.

Touchtone banking, intelligent text-to-voice conversion, interactive touch screen monitors, and digital video are all part of this exciting technology.

Did Bell, whose interest in sound was sparked by his experiences as a teacher of the deaf, envision the exploding world of telecommunications? No one knows, but one thing is certain. Telecommunications is big, and it is getting bigger every day.

HOW BIG IS BIG?

Consider the international marketplace: $13 billion in 1994, it will grow to $26 billion by the year 2000.

In 1994, only 2.5 billion of the world's 6 billion people had access to a telephone. That year, the worldwide average was only ten telephones per 100 people.

There were 600 million telephone numbers in the world in 1994, including business phone numbers. But over 1 billion more people will have a telephone by the year 2000. In 1993 alone, 28 million new phone lines were hooked up worldwide.

As trade barriers fall, opening new markets abroad, and more advanced telecommunications technologies become accessible worldwide, business's use of international long distance is growing 15 percent per year. Bert C. Roberts, Jr., chairman and chief executive officer, MCI Communications Corporation, says the data communications market alone is growing so fast that it will surpass voice for business customers before 2000. Other spectacular growth is expected in wireless services and for interactive multimedia—an opportunity brought about by the merging of telephones, computers, and entertainment.

MCI Communications Corporation, which compiled these statistics, says that 300 million people will be using the Internet by 2000.

TELECOMMUNICATIONS: CHANGING THE WAY WE WORK

Most analysts who study the telecommunications industry believe "the wired society"—as *Fortune* describes it—will forever change the way businesses operate. Laura Tyson, chief economic advisor to President Clinton, says the United States's comparative advantage is moving in the direction of technology and skills. Cornell University's Stephen R. Barley argues that telecommunications technology will reshape the workplace because of our growing knowledge of how to convert electronic and mechanical impulses into digitally encoded information—and back again—and how to transmit such information across vast distances.

Other researchers have found that although the typical company in their ten-year study eliminated 20 percent of its work force, it tripled its investment in information technology.

Under former command-and-control rules—still the norm in some traditional companies—employees worked under a hierarchical system, reporting to supervisors or bosses who assigned and monitored tasks. Today's networks, however, facilitate teamwork. When information can be shared instantaneously, regardless of where users are, functional teams form, disband, and form again. They get the job done—and telecommunications makes their achievements possible.

Forward-thinking companies are using the new technology to overhaul their traditional operations. One Osaka, Japan-based industrial firm has scrapped its centrally controlled, top-down management structure. Instead, the company's divisions are recognized as independent profit centers and make their own divisions—but they are connected on a corporate telecommunications network.

For businesses that can recognize the new connectivity and the strategic advantages it offers, the payoff is great. For instance, Union Pacific has planned its cost-effective approach to managing the rail business around automation and telecommunications. In 1986, the railroad had 40 offices handling its paperwork; by 1994, two-thirds of its communications with clients were handled through electronic data exchange, centered at a single office.

TELECOMMUNICATIONS EQUIPMENT

The United States leads in the world telecommunications market. As of 1994, the industry served more than 90 million households and 21 million businesses nationwide, with revenues of $193 billion.

Many products fall under the generic heading of "telecommunications equipment." Among them are cellular radiotelephone systems, customer premises equipment, fiber optics, microwave radio systems, network equipment, satellite communications systems, search and navigation equipment, and wireless personal communications systems.

However, companies that once made stand-alone telephone equipment are diversifying and moving into high-end, value-added markets—like Tellabs' Titan 5500 digital cross-connect system, used for managing high-volume digital information. One Titan system, if it were dedicated solely to voice traffic, could transport approximately 700,000 simultaneous telephone calls.

Also included under the telecommunications category are telecommunication services: cellular and radio services, satellite services, and telephone services. Newer, more sophisticated services demand costlier, more complicated equipment. That need translates into products for telecommunications—products you may help make, sell, or use.

YOUR FUTURE IN TELECOMMUNICATIONS

In 1983, when *Opportunities in Telecommunications Careers* was first published, most of the jobs in telecommunications were with vendors of telephone and telephone-related equipment. While there are still many opportunities in vendor companies, there is a rapidly growing need for expertise by users—companies whose management strategies depend on telecommunications and on analyzing the information these systems bring.

Perhaps you will become a network engineer or a data communication systems analyst, a microwave engineer or a netware programmer. Your future may lie in marketing telecommunications services or sales on the Internet or digital video—or even in a job so new it hasn't been invented yet.

In short, the outlook for your employment in telecommunications is bright and getting brighter. Many people have found telecommunications is an exciting, challenging field. Maybe it will be that way for you.

CHAPTER 2

THE INTERNET

If you want to work in telecommunications, or if you're even considering whether you want to work in telecommunications, you need to get on the Information Highway—now.

More and more users are signing on to Internet, the network of thousands of computer networks that links the globe, joining together users from every continent and growing at an estimated 10 to 15 percent per month. San Francisco writer Rob Wood, who authors a regular column distributed electronically on the Internet, calls the network "the Communication Renaissance." Author Steve Lambert describes communication on the Internet as "controlled through a complex set of rules, expressed and enforced through software that runs on computers that are linked together with wires, fiber optic cables, microwave, and satellite communication links."

No one knows for sure just how many users there are. By early 1995, says MCI, over 3.8 million computers were linked to the Internet, with at least 25 million people able to be connected on-line in 154 countries. The Internet has spilled out of the academic world to offer both access to information and a fast, inexpensive means of communication to the general public. As people learn about the Internet and as access becomes cheaper and easier, that number will grow. MCI says another network is connected every 15 minutes. Every month, another 160,000 people become Internet users.

HISTORY OF THE INTERNET

The Internet was born out of the need for people to share information with each other. In one form or another, it has been around for nearly 30 years, when several mainframe computers were networked together to share data. At first, the resulting network, called ARPAnet, was used for national defense. Later, it became the basis of the National Science Foundation Network, NSFNET. At first, most users were government agencies and leading universities. Scientists and researchers used the network and its capability for electronic mail (e-mail, in today's terminology) to exchange information and files.

Initially, computers that used the network were large mainframes. But companies such as Apple, Commodore, and Radio Shack started to produce personal computers (PCs)—single-user machines that nonexperts could easily learn to use. By the 1980s, IBM had entered the personal computer field. Sensing a potential growth market in the increasing acceptance of computers, competitors began to produce IBM-compatible PCs.

Rapid developments in technology allowed manufacturers to offer personal computers with more memory, more features, and more potential uses. Prices dropped, and computers became commonplace—not only for business and industry but also in homes. People began to use their personal computers and terminals to send and receive information over telephone lines.

The telecommunications link that makes this possible is called a modem. This device is either built into the computer or plugs into it. The name comes from the two-way conversion process known as modulation-demodulation, in which the modem converts digital signals from the computer into analog signals that are transmitted over the phone lines and reconverted to digital form. A number of rules and conventions, called protocols, have been set up to make sure that the signals between the two computers are able to match.

WHAT CAN YOU DO WITH THE INTERNET?

Most activities on the Internet fall into four broad categories: sending and receiving electronic mail (e-mail), transferring files between computers, taking part in discussion groups through mailing lists and Usenet News, and searching for information.

You can go to school on the Internet, sending your homework in by e-mail and downloading assignments. You can even earn graduate-level degrees.

Florida's Walden University, regionally accredited, allows active professionals to earn the master of science degree in Educational Change and Technology Innovation. The doctor of philosophy (Ph.D.) degree is offered in four tracks: administration/management, education, health services, and human services, including professional psychology. Walden also offers a doctor of education (Ed.D.) degree.

Most students complete the master's program in two years, but those taking part in the doctoral program usually set three years as a reasonable goal.

For information, write Director, Student Recruitment, Walden University, 801 Anchor Road Drive, Naples, Florida 33940, or contact **request@walden.edu** and leave your postal address.

DOING BUSINESS THROUGH THE INTERNET

Here are stories of four users whose business careers are built around telecommunications on the Information Superhighway.

Computer Services

Gary Hutchins, owner and president of Pro Systems, Inc., a computer company in Charlotte, North Carolina, provides "complete solutions" to 2,000 customers across the Southeast. Some are certified public accountants (CPAs); others are lawyers. Still others are corporations. Hutchins's company provides hardware, networking, software, installation, training, support, wiring, and supplies.

Pro Systems, Inc., also sells long-distance services to more than 100 companies as a dealer for BTI, one of the United States's 15 largest long-distance carriers. Hutchins also offers point-to-point leased lines, calling cards, beepers, and international calling services. He plans to expand into voicemail systems and video conferencing.

Since 1994, he has been teaching Internet courses—on line. Topics include establishing a business presence on the Net; interfaces, power tools, and connectivity strategies; and Internet security. In addition to the courses, Pro Systems, Inc., offers consulting and training, and, if desired, designs and sets up home pages or storefronts on the World Wide Web.

Journalism

Michael OReilly, Canadian freelance journalist, calls the Internet an invaluable resource because it brings people and information together. When an editor assigned him stories about satellites, he researched the information through the Net. Using indexes, he tracked down several satellite-oriented newsgroups and mailing lists. "Half an hour of reading brought me up to speed on the technology and politics of the subject," he says.

Next, OReilly "went" to the Industry Canada site in Ottawa that houses government documents, searching for decisions an agency had made on satellite programming. Then he used a general indexing service to search the Internet for databases maintained by satellite enthusiasts.

To flesh out his story, he asked the discussion groups several questions, inviting users to respond. He interviewed several by e-mail. Finally, he filed his story electronically and used e-mail to complete the final edits.

Technical Assistance

Jim Reiss, computer programmer and a system manager on Unix-based computers, says the Internet plays an ever-increasing role in his job. When he is providing technical assistance with one of his company's products, e-mail lets customers send samples of program code for Reiss to try on his system. Then he can send a patch to customers without needing to use a tape or a disk.

If Reiss is on the road, his co-workers e-mail him questions and information he can access from the customer site. He has even used e-mail to quickly proofread the draft of a book, written in collaboration with a computer vendor.

Consulting

Dan Brinegar, consultant and information developer in Phoenix, says that by the year 2000, "Every job that counts is going to rely on tele-communications as a major enabling tool. There will still be accountants and machinists and lawyers, mechanics and customer-service reps—but they're all going to be on-line."

Brinegar and his father run a consulting operation that melds technical writing, information management, and performance support systems. All focus on the Internet.

As a consultant, he is currently writing or developing electronic manuals and help systems, providing "usability input for a worldwide inter-networked client company," and working with another writer, some engineers, and a manager towards getting documentation on-line over the World Wide Web.

The team develops the reference and task-oriented documentation the client needs to use the network—not only in the United States but at the client's sites worldwide. "I interview these engineers or sit in labs with them to find out what steps are needed to do any particular task involved in running the net or adding new equipment," Brinegar explains. "Then I write and format it so that any network operator or analyst with standard training can read that document and perform the task.

"I have an international audience. I have to ensure I meet the needs of some audiences in Japan, for instance, who want all there is to know about the operation and the product before they'll do anything, and some, like Germans and Americans, who don't want their information to go into too much detail. It's a balancing act, made more difficult by rapidly changing technology and a loose use of jargon because so many different versions of 'Business English' are spoken."

Brinegar's prediction: by 2000, everything businesses do will be on-line, distributed, and interconnected. "It's the only way to get things done fast enough," he says, "when your customers and suppliers can be anywhere on the planet."

WORLD WIDE WEB

There are several reasons being on the Internet will help you in tele-communications. One of the most important is the World Wide Web, fastest growing area of the Internet, that contains a great deal of infor-mation you can access from your own personal computer. It's like hav-ing the world's libraries in your room.

The Web, as it is commonly called, uses a connection called hypertext that lets you jump from one document on the Web to another. It is a way of displaying information on the Internet that allows readers to skim through material, picking and choosing topics that interest them without having to close one document and open another.

In order to navigate the Web, you'll need a Webbrowser—a type of software that can make the connections you want. Often, your Internet service provider has this software available on-line. If not, you can usu-ally access another site that does. If you have the appropriate software and equipment, you can even play audio files through your speakers and see graphic images on your own computer monitor screen.

The Web is a powerful research reference that lets you search for lit-erally thousands of topics. Journalist Rob Wood, who has written a book about surfing the World Wide Web, has compiled lists of Web sites that deal with specific topics. For instance, MARVEL (the Library of Con-gress Machine-Assisted Realization of the Virtual Electronic Library) is an information storage and retrieval system with information on all reg-istered copyrighted material.

Other U.S. government documents are also available on-line. It is possible to retrieve the full text of bills introduced in either chamber of Congress; data from the 1990 census compared with that of the 1980 census; major policy statements by the President of the United States; and recent issues of the *Commerce Business Daily*—a government publication that announces invitations to bid on projects proposed by federal agencies.

Also on the Web: displays by art museums and galleries, including information on treasures and related history. LeLouvreWeb, developed by an art historian, shows a sample of famous paintings, especially those of Impressionists, and offers a mini-tour of Paris.

Perhaps most important for those who want to work in telecommunications is the fact that you can research companies, post your resume, search job listings worldwide, and even send in applications—all on the Internet. Start with the On-Line Career Center, which is located at **ftp.std.com**.

Another good source on the Web is *Career Magazine,* which summarizes job postings from various Internet networks and updates them daily. Jobs can be searched in various ways: by location, by job title, and by skills required. You can find *Career Magazine* at its e-mail address: **http://www.careermag.com/careermag/news/index.html**. In Chapter 12, you'll learn more about using the Internet to find jobs.

HOW TO CONNECT TO THE INTERNET

You can get on the Internet from your home or business easily and inexpensively. All you need is a computer, a modem, and a connection to an Internet service provider. To access the Internet, you can use almost all popular computers, such as an IBM or compatible running DOS or Windows, an Apple Macintosh, or a Unix workstation. You will need communications software, but many service providers will supply that as part of an Internet connection.

There are two basic types of Internet connections: dedicated lines and dial-up lines. Dedicated lines provide the fastest access to the Internet, up to 30 million bits per second, but they are the most expensive. Access to the Internet through dial-up lines is offered by a number of service providers.

Many high schools and colleges provide teachers and students with Internet connections free or at low cost. Public educators in more than 40 states can get easy access to the Internet. For example, Florida gives public school teachers free accounts on a system called the Florida Information Resource Network (FIRN). Orlando sixth-grade teachers John Parmenter and Gracemarie Howland put their students on-line to "talk" with other classes across the United States and even with selected pen pals abroad. Parmenter uses FIRN to download lesson plans from CNN NewsRoom and The Discovery Channel as well as *Cable in the Classroom* listings.

Sixth graders have "read" electronic copies of literature, such as Frederick Douglass's biography, found in several Internet locations. Says Parmenter, "Students have done research in the libraries of some of the world's greatest universities—while sitting in our Orlando classroom."

Parmenter has downloaded software that enabled him to access sound and graphic files, such as NASA's shots of the 1994 comet that hit the planet Jupiter. Sixth graders have also "talked" with math education members at Florida Institute of Technology, who prepared special lessons for the youngsters, and have taken part in MathMagic and Maya Quest (a social studies discovery program), both from Minneapolis. And the school's physical education department is participating—via the Internet—in a track meet with a school in Nebraska.

Free-nets, another way of accessing the Internet through dial-up lines, allow public access to the Internet via computer modem and phone line for anyone who wants it. Since free-nets are often funded by one or more agencies, many of them can be used without charge to individuals. However, because the Internet is becoming more popular, you may have to sign up ahead of time to use a free-net or wait your turn at an access site.

Free-nets already exist in a number of metropolitan areas: Akron, Cleveland, and Columbus in Ohio; Atlanta, Los Angeles, Memphis, Milwaukee, Seattle, and Tucson, for example. If you don't live in one of these areas, you still may be able to use a free-net in a region like the Niagara Peninsula, north Texas, or east-central Illinois (Prairienet). Ask your reference librarians at your nearby public or school library for details. Almost certainly they can help you find out what you need to know to get on the Internet. In some cities and towns, the library itself may have a free-net connection.

WARNING

There are several things you must be aware of before you consider signing on to the Internet.

Because the Internet is worldwide and because it has no real censorship restrictions, some material may be adult-oriented, sexually explicit, and offensive to you. Most isn't, but if you are offended, read disclaimers and log off promptly.

Costs for on-line Internet connections and services vary widely and may be priced differently. It is easy to unknowingly run up a bill for several hundred dollars in connect time. Many services require automatic charge-card billing. Often you won't realize the costs until your monthly statement arrives.

Ask questions first about how charges are computed. Must you pay a minimum sign-on fee, even if you have no messages, every time you look to see? How many hours per month will your provider give you for a set price before surcharges start? How will you know when you've reached or exceeded those hours?

Are there surcharges for signing on through Tymnet or Sprintnet or other required telecommunications gateways during prime time business hours? In 1995, for example, Tymnet was charging Delphi users an additional 15 cents per minute for logging on from 6 A.M. to 6 P.M. However, the surcharges didn't apply to evenings or weekend use.

Are there local access numbers you can use, or must you incur long-distance charges just to connect to the network? Major cities and suburbs usually offer choices; those who live in rural areas may need to place long distance calls. Some access numbers are reserved for higher speed modems; that is, a person who calls an on-line service with a 2400 baud modem may be given a different telephone number to dial from a person who is using a 9600 baud modem or a 14.4 modem.

Is an off-line mail reader program available? Connect time (the time your computer and telephone are physically connected to the network) is costly and can add up fast. If you can download e-mail messages quickly, reading them and responding when you are not on-line, you'll save considerably. Subscribers to various lists may receive as many as 300 pieces of e-mail a day. Reading the messages screen by screen and keying in a response while you're still hooked to the Internet through a telephone line on which charges are accumulating is not healthy for bank accounts.

Some services will sell or recommend an off-line software program that is compatible and that will pull your messages automatically at pre-set times. Then you can read your messages when you're ready, compose your replies leisurely, and upload them to the Net quickly and economically at nonpeak times.

THE INTERNET AND YOUR FUTURE

As Brinegar puts it, "The expansion of commercial, educational, governmental, and personal communication via Internet is expected to contribute to what will become a nearly one-trillion-dollar-a-year industry with hundreds of millions of users worldwide by the end of the decade.

"People entering college now have been on the Internet since junior high school; if you're 20 years old and you haven't been on-line already, you're in big trouble."

Is he right?

No one knows for sure. But if you're going into telecommunications—if you're even thinking about telecommunications as a career—you'd better start cruising the Information Superhighway as quickly as possible. Find an access ramp and be part of the Internet traffic.

CHAPTER 3

DIVESTITURE: THE AT&T BREAKUP

No single event of this century has had as much influence on telecommunications as the breakup of the American Telephone and Telegraph Company in January 1984, the result of a long-running court battle between the U.S. Justice Department and AT&T. Divestiture was responsible for a fundamental restructuring of the entire U.S. domestic telephone business.

Before that date, AT&T had over 20 wholly owned telephone companies. After that date, they were no longer part of AT&T. These wholly owned companies were reorganized into seven regional Bell holding companies whose main business was to provide local phone service.

Author Steve Coll, who read more than 50,000 pages of trial transcripts, depositions, pleadings, internal memos, handwritten notes, and calendar and diary entries dealing with the antitrust suit, says the dismantling of the world's largest corporation cost AT&T $100 billion, or three-quarters of the company's entire assets. In nearly every state between 1982 and 1985, he argues, the price of basic residential telephone service increased significantly, while the quality of service declined. The cost of installing and repairing telephone equipment also rose dramatically as the one-stop shopping telephone customers had enjoyed for a century gave way to a pluralistic new industry.

WHAT LED TO DIVESTITURE?

The events leading to AT&T's divestiture are complex, and they extend over several decades. Many people agree that the 1956 settlement of an antitrust suit (brought by the Department of Justice against AT&T in 1949) effectively shut AT&T out of growth areas such as data processing and other unregulated telephone areas. As a result of the settlement, Bell technology had to be licensed to competitors, and Bell was required to limit itself to the regulated telephone business.

Another significant event occurred in June 1968 when the FCC, in the Carterfone decision, allowed non-Bell equipment not detrimental to the network to be connected to it. The Carterfone decision meant that as long as equipment from non-Bell sources was compatible technically, AT&T could not prevent customers from attaching it to their telephone lines. The eventual effect was to open the customer-premises equipment market to competition.

By 1969, Bell and AT&T were being attacked on a new front. Microwave Communications Inc. (MCI) applied for and received FCC permission to build a private microwave network for long-distance calls between Chicago and St. Louis, offering rates below those of AT&T. By 1972, MCI chairman William G. McGowan combined MCI's original company with 17 additional regional companies, creating a network with radio towers to relay phone messages on microwaves.

MCI then requested the FCC to approve Execunet Service, which would let a customer call any other phone in an area that MCI served. In 1975, the FCC found that MCI's Execunet Service exceeded its operating authority and was an unauthorized version of message toll service. That decision, however, was reversed in July 1977 by a District of Columbia Appeals Court. By April 1978, AT&T was told to offer interconnection facilities to MCI.

Meanwhile, McGowan and MCI filed an antitrust suit, settled in June 1980 by a $1.8 billion damage award against AT&T, which was promptly appealed.

While all this was happening, the U.S. government filed an antitrust suit against AT&T, charging it with monopolizing interstate commerce in communications. Although AT&T wanted the case dismissed, arguing that only the FCC had the right to regulate AT&T, the U.S. Supreme Court eventually accepted the government's argument that phone industry competition issues should be settled in antitrust court.

In 1978, Judge Harold H. Greene inherited the antitrust case from Judge Joseph C. Waddy. After much legal maneuvering and intense negotiations, chronicled extensively by Steve Coll in *The Deal of the Century: The Breakup of AT&T,* it was Greene, eventually, who would approve and enter the Modified Final Judgment that made divestiture effective on January 1, 1984.

AFTER DIVESTITURE

As part of divestiture, AT&T spun off the seven regional holding companies. AT&T kept Western Electric, Long Lines, the embedded base of technical equipment, and the assets to provide interexchange long distance services. Although customers got their choice of carriers in the equal access process, AT&T was able to keep over 80 percent of the interstate long distance market. Meanwhile, the regional holding companies received various waivers that allowed them to enter the markets for computers and software, cellular telephones, consulting, equipment leasing, and foreign business ventures.

In 1987, Judge Green removed the restrictions that kept the regional holding companies from nontelecommunications businesses. Court rulings and legislative action later clarified a number of issues.

By 1992, the Federal Communications Commission had voted to expand competition in local telephone service. The FCC allowed smaller communications companies to sharply expand their use of networks operated by regional Bell companies and other local phone carriers.

Although Congress debated a sweeping overhaul of the nation's telecommunications laws in 1994—a change that basically would ease restrictions on telephone companies if they gave up their traditional monopolies, the U.S. Senate adjourned without finalizing legislation.

WHAT IS AHEAD—AND HOW IT AFFECTS YOU

If you are interested in a telecommunications career, you will want to monitor the legislative and regulatory environment in which telecommunications companies are forced to operate. That is because these decisions affect what the companies are allowed to do; that, in turn, impacts the technology and number of jobs available.

For instance, the 1994 FCC-sponsored auction for the right to use the public airways initially awarded ten nationwide licenses; these were expected to be used for advanced two-way paging services. Consequently, Economic and Management Consultants International, a market research organization, estimates that U.S. pager sales will grow to 40.5 million users in 1998, while the world market will grow to 111 million from 40 million.

The United States Telephone Association, national trade association of the local exchange carrier industry, believes that changes in public telecommunications policy are necessary if American businesses are to compete effectively in the growing global marketplace.

Specifically, USTA supports:

- maintaining universal service
- building advanced network capability
- providing a seamless nationwide network
- assuring an adequate telecommunications infrastructure to meet the needs of public health, safety, defense, and security.

How quickly telecommunications policies and regulations are changed—and to what extent—may determine your opportunities and affect your choices as you plan your career.

CHAPTER 4

RESEARCH AND DEVELOPMENT

Because telecommunications technology is developing so rapidly, there are many opportunities for people who are interested in research and development. Computers and communications are especially promising fields; video and television must be added as the converging technologies produce new opportunities. For instance, it is now possible for cable operators to bring combined cable TV and telephony services to business and home business over hybrid fiber/coaxial networks. As cable companies become full-service providers, business and residential customers can soon access not only basic telephone service but also on-line services, the Internet, and additional future on-demand applications. In short, the same coaxial cables that bring TV into the home can also carry phone service and other two-way communications. One of the United States's largest cable multiple system operators is already testing delivery of telecommunication services to businesses and homes in greater Syracuse, New York.

However, at the same time as technology advances make new applications possible, they may lead to unanticipated problems. The National Telecommunications and Information Agency (NTIA) warns that today we may be aware of only a small part of the potential risks involved in global, instantaneous, high-capacity networks that combine voice and data transmissions with computer access and control. Security and integrity of communications are two major issues that must be addressed.

R&D SPENDING

One way to learn just which companies stress research and development is to obtain and study annual financial reports. You can find company addresses by checking with your local public library or by reading trade publications. Often, annual reports may be offered through listings in such publications as *Fortune* or *Business Week.* Or you can write the company directly, asking it to mail you a current report and to put your name on its mailing list for future reports.

Using such a report, for instance, would let you know that Tellabs, Inc., a global supplier to the communications industry, spent almost $65 million on internal product development in 1994 and $80 million in 1995—much of it on enhancements to its digital cross-connect systems for managing high-volume digital information.

You can also search the World Wide Web of the Internet for company sites, looking for telecommunications-related corporations and information about their research and development programs. For instance, the command **http://www.bst.bls.com/bbs/betelecom.html** lets you know that BellSouth Telecommunications Inc., with headquarters in Atlanta, Georgia, serves over 19 million local telephone lines. The company does business as Southern Bell in North Carolina, South Carolina, Georgia, and Florida and as South Central Bell in Kentucky, Tennessee, Alabama, Mississippi, and Louisiana. Among the leaders in optical fiber deployment, Southern Bell and South Central Bell have over 1 million miles of this high capacity transmission medium in their network.

BellSouth Telecommunications Inc. carries on intensive theoretical and practical research in telecommunication network and instruments, data networks, and man-machine relations. Projects concentrate on the design, organization, economic, and management aspects of telecommunication systems.

The company's studies involve problems of speech recognition, speech identification, and speech production by machine—research aimed at producing machine intelligence that can exchange information. Speech-processing products are based on the company's previous research in eight text-to-speech converter languages: modern Arabic, Dutch, Esperanto, Finnish, German, Hungarian, Italian, and Spanish.

networkMCI DEVELOPERS LAB

In 1994, MCI's networkMCI Developers Lab in Richardson, Texas, began operations as the first nationally sanctioned testing and certification facility for advanced telecommunications technologies. Its backer: the North American Telecommunications Association (NATA), the leading national trade association that represents business and public communications suppliers. NATA's more than 700 members include manufacturers, distributors, software developers, and systems integrators who provide a full range of integrated voice, data, and video systems and services. Its purpose is to help companies and integrators develop and bring to market computer-telephone-integration (CTI) applications.

At the laboratory, telecommunications hardware and software developers test their applications on a "live" network. Developers can tap into MCI's network and access everything from narrow-band analog voice services to ultra high-speed digital data services. The laboratory is staffed by technical specialists in voice and data communications. There is also an extensive database that allows developers rapid access to many sources of information.

AT&T BELL LABORATORIES

Mention telecommunications research, and most people immediately think of Bell Laboratories. That is because Bell Labs has been the giant in telephone research. Members of its technical staff have received over 20,000 patents since the labs were founded. In fact, AT&T Bell Laboratories scientists and engineers have been awarded an average of more than one patent per working day since 1925, when AT&T Bell Laboratories was incorporated.

The divestiture of AT&T in 1984, however, resulted in changes. Although AT&T got to keep the labs, a number of employees were transferred to other parts of the company.

Today, AT&T Bell Laboratories, the research and development unit of AT&T, employs people in six states. Its mission is to provide basic and applied research, development and design, systems engineering, and information and operations systems for AT&T.

Seven AT&T Bell Laboratories scientists have received Nobel prizes: for discovery of the wave nature of matter, for discovery of the transistor effect, for theoretical contributions on the electronic structure of magnetic and noncrystalline materials, and for discovery of cosmic background radiation.

Laser research, information theory, lightwave communications, electronic switching, and communications via satellite are among the many areas in which AT&T Bell Laboratories scientists and engineers have made major contributions to science.

AT&T also has an engineering research center, located in Princeton, New Jersey, which provides manufacturing process research and development for AT&T. Scientists and engineers work on a variety of leading-edge technologies, including advanced automation, light energy, semiconductor processes, engineering computer aids, materials processing, and interconnection technologies.

SPECIAL AT&T BELL LABORATORIES PROGRAMS

The Laboratories offer an engineering scholarship program that helps outstanding minority and women high school seniors who have been admitted to full-time computer science, computer engineering, electrical engineering, mechanical engineering, or systems engineering studies at accredited four-year colleges or universities. Each year, 10 new scholarships are awarded, and if participants maintain at least a B average, they are renewed for the four years of college. Winners are chosen from approximately 700 to 800 applicants. Deadline for applications and all supporting documents for the following academic year is January 15.

Another AT&T Bell Laboratories program offers minorities and women the opportunity to do summer research. This program is available for undergraduate women who have completed their third year of college. It emphasizes ceramic engineering, chemistry, chemical engineering, communications science, computer science/engineering, electrical engineering, information science, materials science, mathematics, mechanical engineering, operations research, physics, and statistics. Successful candidates work for at least 10 weeks, starting in early June, and are paid at a level equivalent to salaries of regular AT&T Bell Laboratories employees with comparable education. Deadline for applications and supporting documentation for the following summer work term is December 1.

Beginning graduate students who are members of underrepresented minority groups may apply for AT&T Bell Laboratories cooperative research fellowships, which are awarded for graduate work leading to the doctoral degree in chemistry, chemical engineering, communications science, computer science/engineering, electrical engineering, information science, materials science, mathematics, mechanical engineering, operations research, physics, and statistics. Each fellowship is renewable yearly for four years, contingent upon satisfactory progress towards the doctoral degree. Deadline for applications and supporting information for the following academic year is January 15, and all supporting material must be received by January 31.

A graduate research program for women, funded by AT&T Bell Laboratories, is designed to identify and develop research ability in women and to increase their representation in science and engineering. Participation in the program begins at the start of the candidate's graduate education. The program provides financial support for outstanding women students who are pursuing full-time doctoral studies in the following disciplines: chemistry, chemical engineering, communications science, computer science/engineering, electrical engineering, information science, materials science, mathematics, mechanical engineering, operations research, physics, and statistics.

Four fellowships and six grants are awarded annually in early April to women beginning doctoral studies the following September. Applications must be received by January 15, and all supporting material must be received by January 31.

CHAPTER 5

MANUFACTURING AND SELLING

Since divestiture (the breakup of AT&T) in 1984, many changes have occurred in telecommunications manufacturing. Statistics from *NTIA Telecom 2000,* the October 1988 publication of the National Telecommunications and Information Administration, U.S. Department of Commerce, show that in 1980, the United States had a trade surplus exceeding $7.3 billion in electronics-based products. By 1986, however, this had become a deficit of almost $16 billion. Before divestiture, the United States had a significant trade surplus in telephone equipment; by 1988, it had a deficit of over $2.7 billion.

However, even though the U.S. trade deficit in telecommunications had imports exceeding exports by nearly $3.3 billion in 1992, primarily for consumer products, this was offset by a $1.3 billion trade surplus for large switches and network transmission equipment.

Data from the North American Telecommunications Association (NATA), which tracks the telecommunications industry, predict that in 1997, the market for telecommunications equipment will total $103.8 billion—a growth of nearly 15 percent per year from 1992. Data and networking equipment should see the largest expenditures, with $28.5 billion in sales being projected for 1997. Network equipment, with $24.7 billion in expected sales for 1997, is second.

Other market segments and their projected 1997 sales include emerging technology equipment ($9 billion); mobile communications equipment ($6 billion); facsimile (fax) equipment ($8.6 billion); and call/voice processing equipment ($5.9 billion).

The breakup of the Bell system was a significant factor in giving foreign manufacturers easier access to U.S. markets for telephone equipment. Yet U.S. manufacturers did not have the same opportunities overseas. Traditionally, telecommunications in most countries have been controlled by a single government agency that handled postal, telephone, and telegraph functions as a government-owned and operated monopoly.

That situation is beginning to change. In 1996, shares will be sold in Deutsche Bundespost Telekom, one of the giant European companies. And that's just the start. One London investor predicts $55 billion in telecommunications stock gains between 1994 and 1998. However, Europe's $130 billion telephone market—mostly basic calls—won't really open up until 1998, the date when state phone companies start phasing out. Some countries have until 2003 to begin to privatize.

Certainly any change from a government-owned monopoly in telephone service means increased opportunities for U.S.-based telecommunications companies, which are quick to seize the advantage.

For instance, through an alliance with Stentor, an association of Canada's major telephone companies, Tellabs Canada already helps those companies provide uniform, leading-edge products and services across Canada and internationally. Tellabs has also sold its network solutions to leading telecom service providers in Australia, Germany, Great Britain, Hungary, Singapore, Spain, and Sweden.

SELLING SERVICES

Forming joint ventures and entering agreements with foreign companies are ways in which U.S. vendors hope to expand services—and, of course, revenues. In 1993, MCI and British Telecommunications announced an alliance. The joint venture company plans to invest $1 billion to develop a full suite of global communications services aimed at the fast-growing $10 billion multinational businesses market. Initially, the alliance planned to provide more than 5,000 "access points" to multinational corporate customers in 55 countries.

In 1994, MCI announced a new alliance with Grupo Financiero Banamex-Accival (Banacci), offering MCI-like services in the Mexican market. Banacci, the largest financial services holding company, operates an extensive private line and VSAT network that connects 250 cities in Mexico for point-to-point voice and data services. MCI is enhancing this network with fiber-optic cables that connect Mexico City, Guadalajara, and Monterrey.

And Delphi Internet Services, a provider of Internet access, expanded globally to include UK local access. Members in the UK now can receive headline news and technology, marketing, education, and other information from the *Times of London.*

In addition, UK users of Delphi Internet Limited can receive other on-line service news, classifieds, and on-line shopping and can take part in more than 70 custom forums and discussion groups (including an employment forum for tips on job hunting in the UK).

VALUE-ADDED SERVICES

In the years ahead, plain-old-telephone-service (called POTS by the industry) will never be the same. *U.S. Industrial Outlook* predicts a variety of specialized offerings that can be accessed over the regular telephone network or via special carrier networks. Services like audio-conferencing are already in place; what's new includes digital video—one of the hottest technologies on the market.

The ability to transform the moving image into digital terms opens up tomorrow's interactive multimedia network. That may translate into a telecommunications job for you.

Telephone companies are beginning to partner with media, cable, and entertainment providers. For instance, in 1995, MCI joined Horizon Cablevision of Michigan in testing an automated distribution system for cable advertising. Under this plan, cable companies will be able to dial up and pull ads directly off the MCI network rather than having to rely on overnight tape delivery from ad agencies. And, through an automated "spot inventory" for advertisers, ad agencies can dial up, check on all the "avails" through an entire region, and then book the ads on-line.

Says Bert C. Roberts Jr., MCI's chairman and chief executive officer, "The cable industry needs a national network—a way to perform accurate, usage-based billing and an architecture that permits two-way communications. MCI offers a national unifying network that connects cable head-ends through MCI metro fiber rings. MCI's back office support systems can be easily applied to video—on demand."

By 1995, the cable industry already had broadband coaxial cables that extended into 60 million American homes. In future scenarios that Roberts envisions, MCI's long distance network and local fiber rings, combined with cable facilities, will be a mix that competes with the Bells. Roberts also sees MCI's wireless technology used for program selection devices that tune in to cable stations.

Other providers are also jumping into the interactive multimedia market.

Delphi Internet Services Corporation has technology in place that plugs consumers directly into entertainment offerings. By 1995, Delphi users could download photos and bios of television and film stars or could challenge subscribers around the world in competing real-time contests. Technology that combined detailed graphics with vivid sound effects let participants "play" games from aerial combat to intergalactic travel.

Because Delphi is owned by Rupert Murdoch's News Corporation, the company has access to movies and other materials from News Corporation's Fox Broadcasting Company and Twentieth Century Fox and plans to use those "libraries" to develop and deliver multimedia offerings.

No one is sure just what these developments will mean in the years ahead. However, a new multimedia industry is being created as the technologies of communications, computers, consumer electronics, and entertainment converge. Telecommunications companies are exploring business opportunities into this potentially lucrative market—a market that may mean a telecommunications job for you.

CHAPTER 6

TELECOMMUNICATIONS SERVICES

New technologies and their applications have opened opportunities for those seeking careers in the exploding field of telecommunications. Here are three areas in telecommunications services that offer challenge and growth.

VOICE RESPONSE

Voice response equipment and voice mail are not identical. Most of us have used voice response equipment when we've dialed our bank, keyed in our account number on a push-button phone, and heard directions on what buttons to push in order to check our balance, find if a check has cleared, or verify deposits.

Voice mail, however, can be thought of in two ways: as an answering machine on which the caller can leave messages and as a technology that lets callers record a voice memo, address it, and send it simultaneously to a number of people. This time-independent service doesn't require respondents to be at their desks. It provides faster communication than do written documents, and it allows people to respond without having to dictate, sign, and mail letters. Users can retrieve their messages from any touch-tone phone in the world, regardless of time zones.

Voice mail works within companies, too. An executive or sales manager can send a voice mail message to the entire staff or to selected lists, reaching them immediately with instructions or announcements.

Karen Varnas, AT&T division manager for PBX peripherals, voice processing, and voice messaging products, heads product management for this technology. "My responsibilities are like managing your own small business," she says.

"You must understand market and customer needs, translate those needs into product features and functions, and work with research and development to get the product designed." Varnas also has similar responsibilities for PBX data equipment, which allow the PBX to be used for data switching and call management—the system that distributes calls coming in to a PBX to various agents and generates real-time management reports on how calls are being handled.

"You can sit there and watch the screen, knowing that one group of agents is fully busy with calls waiting while another group is idle 40 percent of the time," she explains. "Based on that information, you can modify your call management software to minimize the time customers have to wait before someone picks up the phone."

Varnas, whose undergraduate degree in computer science is from Purdue University and who earned an M.B.A. from New York University thinks voice processing technology may be the fastest growing area of communications.

"There'll be plenty of jobs," she predicts, "for those interested and qualified." She recommends aiming at sales, preparing yourself with a business/marketing background. Product design or technical support each requires an engineering degree, while product management jobs usually require a marketing and business orientation with a technical background.

Tops on the list of requirements she thinks companies look for while hiring is the ability to work as part of a team. Communication skills and the ability to make presentations, both oral and written, are also significant to prospective employers.

TELECOMMUNICATIONS COST CONTROL

One of the problems telecommunications managers face is the ability to allocate telecommunications expenses properly. At Tymnet, one of the world's largest and most powerful public packet data communications networks, Randall Chun, manager of telecommunications cost control, heads a department that authorizes and researches all Tymnet's phone bills—internal, as well as for customers.

Chun, who graduated from high school in Hawaii, received his bachelor's degree in economics from the University of California at Santa Clara and his master's degree in economics from San Jose State University. Most of his work for Tymnet combines economic analysis and cost analysis.

"My job is exciting," he says. "It's one of the best jobs I've ever had because you get things done. You build an incredible amount of excitement into people. They are aware not only of what is going on in their own job but also the reason for what they're doing.

"When you're in a group called Telecommunications Cost Control, you know you have a significant impact on the company's bottom line. Expenditures for phone bills are the single largest budgetary item in the company. So if you can reduce them, you affect the company's bottom line. You know what you are doing makes a difference."

Chun describes his job as controlling telecommunication costs, doing everything possible to keep them down. His department makes sure Tymnet pays the correct amount on bills. "We identify projects which will result in new services or projects to reduce loss," he explains. "We're always looking for a cheaper way to do things. We work with vendors in a variety of ways to lobby for new services or to negotiate for services and get their price down.

"The problems we work on involve a variety of technical, management, and people-related issues," Chun explains. "My job is to perceive what Tymnet's needs are."

Then his group identifies what has to be done and establishes timetables and priorities. Project planning involves things done within

Chun's group as well as tasks done across group lines, involving development, marketing, sales, and finance.

Chun's group has been responsible for reducing costs approximately $2 million a year for the past several years.

"Part of my job that's really exciting," he says, "is melding databases, using the power of the computer to give people the leverage they need...understanding the tools of technology and database management in PCs and mainframes.

"I look for key people who can take what you ask them to do, do it, and go beyond what you even thought about doing," Chun says.

PHONE SECRETARIES

Varnas and Chun, each with a master's degree, have used their knowledge and experience to advance within their respective companies. But what opportunities exist for persons who would like to work in telecommunications and are just beginning their careers? Or for those returning to the work force after being at home for several years?

One possible position is that of phone secretary. That is the term a Chicago-area company uses to describe the men and women who work in its nine city and suburban offices, handling its broad spectrum of answering services.

"We take messages for clients and relay them," says General Telecommunications's Esther Sutrick. "We make appointments for doctors and take claims for insurance companies. Sometimes when you see an 800 number, we are that number, taking orders.

"We provide wake-up service when asked and locate maintenance personnel who are on 24-hour call for apartment complexes when there's no hot water or someone's stuck in the elevator."

The company handles calls to Alcoholics Anonymous, a rape hotline, and a special drug enforcement line for law enforcement personnel.

Technology changes have made handling subscribers' calls simpler. An individual or business wanting the phone secretary to take over uses a touch-tone phone to enter certain codes, locking on-call forwarding to the appropriate General Telecommunications office. When the client gets a call, the phone secretary sees on the computer screen the incoming call, the number of rings before the phone is answered, and the client's name and hours of service, even before he or she picks up the receiver. If necessary, the phone secretary can flip to the next computer screen, showing the names and titles of company officials for whom General Telecommunications takes messages and any special instructions that should be relayed to callers.

Messages are stored in the computer and are given later to subscribers who phone in with identifying passwords, although one customer has a special device on her office desk that prints the messages out as soon as they are received.

Phone secretaries, who used to use plug boards and cords to answer phones, were trained for a month on computer equipment brought to the General Telecommunications office, says Sutrick. "At first they were scared since many of them didn't type. Now they all do."

Since the General Telecommunications office where Sutrick works is open 24 hours a day, phone secretaries work in shifts and are hired for full-time or part-time work. Staffing depends on the time of day (mornings are heaviest), and a number of working mothers choose part-time jobs. In the summer, college students home for vacation will often work 4 to 8 P.M.

"When we hire," says Sutrick, "we look for a pleasant voice, a good telephone tone, spelling skills, a good memory, and the ability to speak English that everyone can understand. Now we are looking for computer or typing skills as well.

"Being dependable counts; so does good attendance and the ability to have patience with demanding callers," she says. Also important is the ability to work with others and to see where help is needed. "If you're free and the person next to you has three calls coming in, you're expected to pick up one."

During the hiring process, applicants hear tapes of incoming calls, answer them, and are recorded. Then tapes are played back so applicants can hear how they sound on the phone. A spelling test, an eye-contact test to match numbers, and a reverse memory test come before the personal interview with the office manager. "After they've read a story we handed them," says Sutrick, "they're asked questions. What was the name of the company president? Who did the caller ask for?"

Successful applicants are put on the payroll and then trained. After three months, they are interviewed again. If it looks like they're staying, they get a raise; subsequent raises come at least once a year and are tied to responsibilities and skills.

Sutrick, who moved up from phone secretary to inside sales, loves her expanded responsibilities, though she still spends two hours a day answering phones. At 9 A.M. she switches to the sales role, visiting with people who come into the office wanting to buy or lease equipment or answering phone inquiries. Before she started selling, she received three months of training to familiarize herself with pagers, fax machines, answering machines, and other equipment General Telecommunications handles.

When a prospect responds to the company's advertising, Sutrick answers their questions, finding out their needs and explaining various models that can help. If they are looking for a new phone system for their home or office, Sutrick can—and does—sell it.

Sutrick is fascinated by the technology, especially with the new alphanumeric pagers that operate within a 75–90 mile radius, displaying an entire typewritten message when the customer pushes a button on the pager in response to the "beep."

Her position is a stepping-stone, she thinks, to becoming a supervisor and, later, a manager. "You can go to the top if you want to do it and work for it," she says. "It's an excellent opportunity for anyone with ambition."

CHAPTER 7

TELEMARKETING

Jennifer liked the rugby shirt in the catalog. She dialed the toll-free number. Within moments, she had given her order to the company representative, billed it to a credit card, and instructed the company to ship the shirt to her office. Jennifer—and millions of people like her—are part of the reason telemarketing has expanded to a $90 billion industry that is still growing.

TELEMARKETING STRATEGIES

Most telemarketing professionals consider telemarketing a marketing discipline that uses remote selling and services techniques to execute a marketing strategy. In short, it is far more than an 800 number or a bank of operators waiting to take your order. Used properly, telemarketing becomes a driving force in a company's strategic planning.

AT&T's Shirley Lanier, staff manager of telemarketing, sees telemarketing as divided into five programs. Each of them, she says, can be thought of as having two strategies: efficient execution and value-added.

Order processing is what Jennifer found when she dialed an 800 number and ordered the rugby shirt. As a company sees it, many of its customers already know its products. Some are repeat purchasers. Others, even though it is the first time they have ordered, are knowledgeable about the products or have a catalog. When they are ready to buy, they need a quick way to order. Individual customers like Jennifer don't want to take the time to drive to a store and buy the rugby shirt from a clerk. For them, picking up the phone is faster and more efficient.

Businesses, too, find ordering through telemarketing saves time. Printer ribbons, computer paper, or other consumable supplies continually need replacing. The staff member responsible for making sure the office is supplied with envelopes, for instance, does not want to run to the local stationery store when supplies run low and certainly will not want to wait until a salesperson physically calls to take the order. Using the order processing services of a reputable company is a lot simpler.

"What you need is a way to buy and a brief buying window," Shirley Lanier explains. "The customer wants to order widgets now, as easily as possible. Telemarketing makes it happen."

If Jennifer, dialing to order the rugby shirt, reaches a company that considers only the efficient execution strategy of order processing, her conversation will be brief. The telemarketing representative will take down the information, keying it in on a computer which may already have been programmed to indicate the shirt's available choices for color and size when the catalog order number is entered. Jennifer's charge card number and desired shipping address are entered, and the transaction is complete. AT&T reports a typical telemarketing order processing call lasts 1.5 to 2 minutes, and that length of call holds across all industries.

If, however, the company Jennifer calls uses a value-added strategy for order processing, Jennifer may be spending three to five minutes with the telemarketing representative. Jennifer may be asked if she wants matching jeans or a jacket. One major clothing-by-phone-order company's representatives have information at their fingertips—literally, since it comes up on screen when they hit a computer key—on sizes and colors to aid customers in making their decisions.

Value-added strategy makes even more sense for business-to-business telemarketing. The philosophy behind it says that even though customers are ready to order a company's products, they may not realize all the add-on items they could be buying that would help them use the products in the best way.

A company that orders a computer through telemarketing, for instance, will probably be asked about printers. A company that orders a printer may be asked about ribbons and paper or laser toner cartridges. AT&T says surveys show callers like this type of information and dialogue because it gives them a chance to consult with the telemarketing representatives.

However, from the selling company's point of view, using value-added order processing has certain drawbacks. If each phone call from a customer lasts twice as long as taking an efficient execution order, it costs the selling company more money because it will need to hire more telemarketing representatives to handle the same volume of calls. There is a break-even point when companies can successfully offer the extra service. On printer ribbons alone, unless the volume of the order is substantial, spending the extra time on the phone call may not be cost efficient for the selling company. If, however, taking the time to interact with the customer produces an add-on sale worth substantial dollars, then offering value-added services becomes profitable.

Customer service is another widely used telemarketing strategy. As AT&T defines the term, it means "support after the sale." After they have purchased a product, customers often need to talk with someone for more information.

Companies that sell consumer products often set up customer service centers to field complaints or pass on additional information. If you buy a package of eight quick crescent dinner rolls, for example, you are given an 800 toll-free number to call if you have questions or comments, want nutrition information, or are not satisfied with the quality of the product.

Companies that use telemarketing as an efficient execution strategy merely accept information as it comes in. AT&T's Shirley Lanier points out, "representatives have a specific set of questions they ask. They merely collect the information. Depending on the company and the kind of product or service, they may give you a time-response commitment.

Then the representative hands your question off to a more experienced person." In a value-added customer service center, however, representatives are expected to take your call, learn your problem, and handle it through its resolution.

INBOUND AND OUTBOUND TELEMARKETING

Order processing and customer service are inbound telemarketing strategies. This means that people who want to reach companies call special 800 numbers. In fact, as early as 1988, Americans dialed AT&T 800 numbers over 6 billion times—more than 16 million times a day. AT&T estimates the number of calls is increasing by a billion calls a year. That means more telemarketing jobs. Both consumer and business editions of AT&T's toll-free 800 directory can be purchased at AT&T phone centers.

Outbound telemarketing, in which telemarketing representatives dial out rather than accepting incoming calls, also is growing rapidly.

The sales support function can be both an inbound and an outbound telemarketing service, depending on company strategy. Generally, however, the term refers to an inside sales group that supports a sales channel. Telemarketing representatives usually do not own the account but support an account executive. For instance, if a hotel or resort sales staff member books a 500-person convention, he or she turns that meeting over to someone in sales support. It is that person who becomes the contact for the convention and works out menus, makes sure audiovisual equipment is reserved, and handles all support details.

In value-added sales support, telecommunications (phones, conference calling, and fax machines) help the account executive to increase value from the sale. For example, if your company buys a PBX, a lot of equipment can be added after the original sale and before installation. Sales support team members can work with you to plan what you need and can handle all the details necessary to get you installed and running.

Account management is more a high-risk, high-reward type of tele-marketing, in which an account executive owns the account. He or she drives the revenue and is held accountable for it as well as for customer satisfaction.

In efficient execution, an account manager might merely use telecom-munications techniques to make sure the marketplace has customer product updates on a very defined product.

Promotion management is one of the newest areas of telemarketing service. In this strategy, a company uses telecommunications to execute or complement an ad or promotional method. For instance, when Kellogg introduced a new cereal, the company ran a toll-free 800 num-ber at the bottom of its magazine ads. Readers could call in and request a free sample.

Because promotion management is tied to a limited-time offer, a company doesn't usually establish an in-house telemarketing center. In-stead, it is more apt to hire an outside service bureau.

ACD AND ANI

Two important technologies used in modern telemarketing are ACD (automatic call distribution) and ANI (automatic number identification). When too many phone calls reach a particular number, they can be placed on hold and assigned to the next available representative through automatic call distribution. In off-peak periods, calls can be evenly dis-tributed among stations. If callers have waited a long time, overflow ser-vice can automatically shift the calls from a high traffic ACD group to a less congested one. And, with 800 automatic number identification, in-tegrated data/phone terminals show the telemarketing operator not only what phone number the caller is dialing from but also important facts from that database about the caller. In fact, with ANI, telemarketers can access the caller's information before the call is answered. If desired, certain proprietary software can be used to capture name, address, and demographic information on incoming calls, without having the operator key it in.

FINDING OUT ABOUT TELEMARKETING

There are a number of sources for learning about telemarketing. One is the American Telemarketing Association, 444 North Larchmont Boulevard, Suite 200, Los Angeles, California 90004. The ATA can put you in touch with local chapters in your area. Often, local meetings, usually featuring speakers, are open to the public for a nonmember fee.

The *ATA Newsletter* is published eight times a year. Recent issues have highlighted critical areas of the telemarketing industry, including selecting the right telephone hardware, voice messaging technology and applications, as well as providing information on customer service. The ATA holds conferences focusing on technology and legislation since various government agencies have guidelines and policies that impact telemarketing. For instance, consumer complaints received by the Federal Trade Commission (FTC) that relate to telemarketing are routed through that agency's nationwide database of companies suspected of telemarketing fraud.

JOBS IN TELEMARKETING

As AT&T puts it, though telemarketing relies on today's most sophisticated communication technologies, beneath it all, it is still people on the phone. Companies that want to make telemarketing work have to recruit representatives, spend time training them properly, keep them motivated while avoiding burnout, and decide how to track and measure performance.

Many different jobs exist in telemarketing. Writing telemarketing scripts, for example, is extremely demanding since a good script is interactive and designed to trigger a specific response.

Telemarketing service representatives are the backbone of the industry. Also significant are supervisors and call center managers. Trainers, service managers, sales directors, and account executives in sales are also needed.

Courses in telemarketing are offered by several schools and colleges. You can contact them directly for information.

Hocking College
   Hocking Parkway
   Nelsonville, OH 45764

New York University
   Continuing Education
   Center for Direct Marketing
   48 Cooper Square
   New York, NY 10003

University of Nebraska at Kearney
   Telecommunications Management
   Business Department
   Kearney, NE 68849-0713

University of Rhode Island
   Marketing
   Kingston, RI 02281

University of Tampa
   Business and Economics
   401 West Kennedy Boulevard
   Tampa, FL 33606

SHOPPING ON THE INTERNET

The term *telemarketing* has traditionally been used to indicate telephone-solicited sales. But a new type of shopping is revolutionizing telecommunications. People are using the Internet to buy and sell products and services.

The Internet Shopping Network is one of the electronic shopping systems available on the worldwide entrant. It is set up like a virtual mall with a variety of stores selling various products. The network offers more than 20,000 computer software and hardware products available from nearly 1,000 different companies. More than 95 percent of these products, representing more than $500 million worth of inventory, are in stock and can be shipped the next business day.

Anyone can browse the stores and catalogs of the shopping network from anywhere on the Internet, using an Internet connection and a copy of a Web browser like Mosaic. The connection address for the network is **http://shop.Internet.net**.

After filling out a membership form and obtaining a member code, you can order products directly from the shopping network. All transactions are processed electronically.

A second shopping mall, internetMCI (offered by MCI Communications), is part of a portfolio of services featuring a new secure electronic shopping mall, a user-friendly software package for easy Internet access, and high-speed network connections to the Internet.

Beta testing for on-line electronic shopping began in 1995. A proprietary digital signature system certifies and identifies valid merchants for internetMCI. The complete system allows customers to shop electronically and make secure transactions directly over the Internet without having to fear that their credit card numbers or other sensitive information will be stolen by electronic eavesdroppers.

"Transaction security is the last major hurdle to making the Internet a viable marketing and distribution channel for businesses," says Timothy F. Price, president of MCI's Business Markets. "By 2000, MCI expects commerce on the Internet will exceed $2 billion and be as common as catalog shopping is today."

BRINGING TECHNOLOGY AND MARKETING TOGETHER

If you are considering telemarketing as a possible career, you should get information on High Tech Direct—a conference that brings together leaders in the high tech marketing industry. There you'll learn about direct mail, research, on-line services, telecommunications technology, the use of fax, direct response TV, CD-ROM marketing, and multimedia.

Co-sponsored by *Advertising Age,* Prodigy, Tandem, Business Marketing, and CMP Technology, this forum for marketing products and services to today's high tech cultures features sessions on the latest marketing developments in electronic, interactive, and print media.

Also highlighted: case studies and user testimonies; direct response TV; radio, on-line, print, CD-ROM, database, and international marketing.

For more information on High Tech Direct 2000, call 800-808-EXPO.

CHAPTER 8

ELECTRONIC MAIL

When freelancer Vicki Gerson wants to talk with one of her fellow writers, she doesn't pick up the telephone. Instead, she goes to her computer, loads her communications software, and dials an 800-access number. A high-pitched tone and a computer monitor screen that suddenly goes blank tell her she is connected to MCI Mail, one of the popular electronic mail services.

Prompts on the screen ask Gerson to enter her user name and a secret password assigned to her by MCI Mail. If she keys in both correctly, she sees the message "connection initiated." A note scrolling rapidly across the screen gives her the day's top headlines, asks if she wants to learn more about them by accessing Dow Jones Information Service, and tells her how many electronic mail messages she has in her in box.

Gerson uses electronic mail as a telecommunications tool that handles her messages regardless of whether she is home. Commonly referred to as "e-mail," the electronic messaging industry is growing rapidly.

"From international trade to sales to publishing, and across corporate America and around the world today, electronic mail technology is being used for a dazzling array of communications requirements," says Michael F. Cavanagh, executive director of Electronic Mail Association, the Washington, D.C.-based trade association that represents more than 160 companies involved in all aspects of the electronic messaging industry. EMA members include systems operators, equipment manufacturers, software developers, industry consultants, and major users. In fact, user members make up almost one-third of the association's membership.

The Electronic Mail Association defines *electronic mail* as the generic term for the noninteractive communication of text, data, images, or voice messages between a sender and designated recipient(s) by systems that use telecommunications links.

Computer-based messaging is the heart of the electronic mail industry. Millions of Americans already use their personal computers and terminals to send text messages to co-workers and friends who have similar systems. Electronic mail gets messages to recipients quickly. What is more important, it negates the geographic distance and differing time zones between the sender and the receiver.

Through enhanced, value-added services—like the ability to create lists and send the same news, such as price changes or product announcements, with only a few keystrokes—electronic mail can reach one or one thousand customers immediately.

SENDING DATA OVER PHONE LINES

The telecommunications link that makes it possible to use personal computers for electronic mail is called a modem. This device is either built into the computer or plugs into it. The name comes from a two-way conversion process known as the "modulation-demodulation," in which the modem converts digital signals from the computer into analog signals that are transmitted over the phone lines and reconverted to digital form.

When you want to send a text file over the phone lines by electronic mail, you key the words into your computer. Many users choose to create the message ahead of time, using word processing software; store the file (either on a floppy disk or a hard drive); and upload the entire file with their communications software, since that process is faster and more error-free than keying in the message on-line.

A number of rules and conventions, referred to as protocols, have been set up to make sure that the signals between the two computers are able to match.

DIGITAL DATA TRANSFER

Coming within the next few years, especially for business communications, is ISDN, which stands for Integrated Services Digital Network. ISDN systems provide simultaneous voice and high-speed data transmission through a single channel to the user's site. A special switch, controlled by a computer in the telephone central office, enables ordinary copper wire pairs to carry the signals. ISDN will let users exchange digital data without having to use modems.

Telecommunications technologies are changing rapidly. That is why every vendor and every country is working out standards. Planning and implementing efficient multivendor networks in a competitive market environment isn't easy when devices depending on complicated electronics must work together.

One set of standards governs specification and protocols for public packet switched networks developed by the Consultative Committee on International Telephony and Telegraphy (CCITT) for the Geneva-based International Telecommunications Union. In this technology, messages are broken down into smaller units called packets. Packets are individually addressed and sent through the network in short bursts at high speeds. Because the intercity phone lines are only used when there is actual data to be transferred, the intercity circuits are not tied up the entire time the caller is on-line. In public packet-switched networks, PC users dial a local telephone number to reach the system through a node junction point.

MARKETING ELECTRONIC MAIL

Sally Baylaender doesn't work directly for an electronic mail network, but marketing MCI Mail is one of her most important jobs. Business Systems Solutions Inc. (BSSi), located in one of Chicago's northern suburbs, is an independent agency that resells MCI Mail and supports its MCI Mail customers. Sally is agency coordinator, a 20 to 30 hour per week responsibility that calls for her to manage the MCI Mail agency. The rest of her time goes to Business Systems Solutions Inc. Digital Information Services Corporation (BSSi DISC), the parent company. She describes herself as a middle person between MCI Mail headquarters in Washington, D.C., and the E-mail's customers.

"Nothing in my background as an English and rhetoric major at the University of Illinois prepared me for what I'm doing," Baylaender says. "I took courses in advertising and typography, gearing myself towards publishing. When I joined BSSi after graduation, I started getting involved with electronic mail. The agency sells hardware and software, and I learned a lot about different types of computer systems. Then, because I'd picked up most of the usage and support and worked with MCI people, I was eventually put in charge of recruiting agents and developing our program."

Before she tackles anything, Baylaender reads her MCI Mail. A particular software program, Lotus Express, automatically dials each of her three mailboxes in turn, downloading as many as 30 messages that have arrived overnight. Some come from customers with questions or problems. Others, usually discussing administrative details, come from MCI Mail headquarters.

Still other messages give Baylaender tips on possible MCI Mail customers, which she will screen and pass on to the appropriate sales personnel. If a current user recommends a prospect, the lead goes to that user's sales agent.

One of the most important parts of her job is making sure agents are paid properly. Since they receive a commission based on the amount of activity in a user account, Baylaender has to calculate the correct amounts and pay them. If agents have major clients, she works with them to develop training procedures and helps run training classes. Training materials are supplied by MCI Mail.

Answering questions from customers takes up much of Baylaender's time. She gets a lot of questions on communications software and on other computer equipment. "I hit print on my screen, but nothing prints to my printer," someone may complain. She's happy when she can suggest time-saving tips to e-mail users. For instance, a small company producing a financial newsletter used to send it overnight to clients. At Baylaender's suggestion, the company put each client on MCI Mail. Now clients can read the newsletter within minutes after it has been sent. The company is happy because e-mail costs are substantially less than overnight air service.

E-MAIL AND THE INTERNET

Until Internet usage exploded, interconnecting the electronic networks wasn't easy. Proprietary technology kept CompuServe subscribers from talking easily to Prodigy subscribers or to America OnLine or Delphi Internet Services subscribers. Suddenly, however, subscribers to each found they could connect through the Internet—and e-mail usage soared.

Between 1987 and 1994, 26.5 million Americans got e-mail addresses, according to *Fortune.* More than 30 percent of all mobile professionals use it. In fact, says *Mobile Office,* one of three computer users spends 20 percent of the time he or she uses a computer to send and receive e-mail messages.

Many of the messages are personal—phone calls, so to speak, made without the customary long distance charges. One of the many pen pal lists to which users of various networks subscribe brings together over 200 persons from around the world, linked together by their ability to share mail messages sent to Penpal-1 or to communicate privately by writing to individual e-mail addresses. Penpal-1 subscribers have learned about Ramadan and Eid, two Muslim observances, from computer programmers in Saudi Arabia; about a teachers' strike in Iceland from a school principal whose classes were affected; and about recipes from Uganda.

Other e-mail lists make it possible for teachers and students worldwide to share information. OWL is an on-line writer's resource developed by Purdue University. Users who send an OWL-request to **owl@sage.cc.purdue.edu** receive automatically (through e-mail) an in-

dex of 80 handouts on specific writing topics, such as the use of com-
mas, overcoming writer's block, and resume-writing tips. Even books
go on-line.

Freelance editor Doug Schmidt's *How to Get Writing Assignments
from Publishers,* a 21-chapter book on nuts-and-bolts successful
freelancing, isn't available in printed form. Instead, someone who wants
to order a copy sends Schmidt a check through the mail. After the check
has cleared, Schmidt uploads the book's text on-line as a single e-mail
document.

Another e-mail source is Newshare, which involves local publishers,
broadcasters, and entrepreneurs in a worldwide system of newsgather-
ing. The high-tech start-up has developed software that will allow news
organizations to put their stories on the Internet, charge readers for
viewing them, track users, and bill them automatically.

MORE THAN JUST MESSAGING

Companies that have spent millions of dollars developing proprietary
e-mail systems, however, may lose some of their marketing advantage if
sending messages back and forth from other systems becomes too easy.
Consequently, several have developed additional services, as well as
lowering prices.

In 1988, for instance, MCI Mail began offering fax service over a
dedicated facsimile network as part of its electronic mail offering. Individual personal computer owners like freelance writer Vicki Gerson can
send fax messages from their own terminals, even though they don't
have a fax board in the computer.

Until this technology was introduced, Gerson had time-consuming
and relatively expensive options if one of her editors at the Chicago
Sun-Times needed a story quickly. Although the newspaper could have a
messenger pick up Gerson's printout at her Northbrook, Illinois, home,
20 miles from the paper offices, the cost would be $38.50, and the time
from the phoned order to delivery at the paper would be nearly five

hours. Or, Overnight Federal Express service would pick up the story by 7 P.M., fly it to the Federal Express Memphis, Tennessee, hub, and deliver it to the *Sun-Times* by 10:30 the next morning at a cost of $14. Instead, Gerson can fax straight from her computer at home, through the MCI Mail gateway, at a cost of 50 cents for the first half-page and 30 cents for subsequent half-pages.

Though Gerson still uses MCI Mail as a backup fax to send copy rapidly and reliably to her editors, today, it is not her first choice. Instead, if an editor needs hard copy (a printout), she'll use the dedicated fax she has installed at her home. More often, though, she'll key in special codes that let the *Sun-Times's* mainframe computer recognize a legitimate incoming message and modem the copy straight to the paper's offices. E-mail carries the message from Gerson's computer into the mainframe, eliminating the need for paper copies.

PUBLIC AND INTERNAL SERVICES

As e-mail networks begin to link, the classifications of "public systems" and "internal systems" will lose their significance. Nevertheless, internal systems users (like Hewlett-Packard) far outnumber those on the public networks. Robert Walker, chief information officer of H-P, says that every month the company's 97,000 employees exchange 20 million e-mail messages and 70,000 more outside the company.

YOU AND ELECTRONIC MAIL

As e-mail services expand, so do job opportunities. Whether systems are public or internal, technicians need to maintain them, marketing experts need to promote their use and explain benefits, and training specialists need to teach users. Maybe one of these jobs can be yours. Soon you will be using electronic mail as a store-and-forward messaging system or as a fast, easy way of exchanging and receiving information.

CHAPTER 9

NETWORKS AND NETWORK MANAGEMENT

Getting computers, modems, printers, and other devices to exchange information with each other is becoming increasingly important in today's fast-paced communications technology. Historically, the term *network* was used to mean the transmission and distribution facilities, including the central office switching apparatus, that comprise the nationwide regulated telephone network. But today, when most people who are not telecommunications professionals talk about networks, they mean groups of computers linked together by cables, phone lines, radio waves, infrared, or fiber optic light guides—at speeds ranging from 300 bytes per second to hundreds of millions of bytes per second. Even the Internet is a loosely linked network of networks.

Fortune reports that in 1994, businesses around the world spent $3.75 billion to buy hardware and software devices that link computer networks.

Some networks run manufacturing systems on the factory floor. The Manufacturing Execution System Association, a trade association that represents software developers and vendors, says industry revenues for software totaled $200 million in 1994. Networks can handle scheduling in plants, can automate payroll information, and can reduce data entry time by as much as 75 percent.

Other networks track inventory and sales, transfer money, handle financial futures, compile and distribute information electronically, or even tie together specialists in various countries who can work together on the network to solve problems for a multinational company. For instance, when Hewlett-Packard engineers are asked for help on a problem a customer reports, the information about the problem and how urgent it is can be entered in a database and sent via the customer-response network to one of 27 centers. Because the database is updated every time an H-P engineer works in a file, each center knows about each job at all times. Hewlett-Packard's networks also handle e-mail. Each month, H-P employees around the world exchange 20 million e-mail messages internally and more outside the company.

UPS's global information network can track a package delivery worldwide, can reroute a driver automatically, can send an image of a signature, or schedule a fleet of airplanes.

With software, networks can even analyze their own performance through remote monitoring, alert network managers before problems arise, and pinpoint and identify the source of trouble.

LOCAL AREA NETWORKS (LANs)

In one of the simplest kinds of LANs, according to computer systems consultant Dan Behm, users share peripherals in an office environment. "Sharon, Jeff, and Kim all work on personal computers," he explains, "but the office has one laser printer. With networking, everyone can use it. Or perhaps one of the PCs has a hard drive with a database stored on it. Networking lets everyone use that hard drive."

A network can have a dedicated file server—one computer that usually handles the network's hard drive and data files as well as other tasks, like printing. Sophisticated network software keeps track of who is using the network and what he or she is handling at any one time.

"For instance," Behm explains, "Jeff and Kim both are keying in order data, using the same database. Each is working on a separate computer, and the file server is yet another machine. Theoretically, four or five or six users can all be typing order records. But the proper kind of software will handle record-locking and file-locking, so if Sharon wants to look at record 25 while Jeff is working on it, the network will not let her get at it."

Another, more sophisticated network may use client-server technology. The server is a powerful, dedicated database "engine," designed to run database questions quickly. Software on an individual's PC (the client) can automatically ask the server for information, and the server sends back the answer. There can be more than one server, Behm explains, so that multitasking can happen. Or, a telecommunications server may have half a dozen or more modems on it, which can be used by anyone in the network. Even video for conferencing or security can be handled by the faster, high bandwidth networks.

Linking Companies Together

A firm like Arthur Andersen and Company, a large accounting firm with offices on several floors in one building, will want to tie its computer equipment together in a local area network (LAN) so it can share resources. But other companies need a wide area network (WAN). A downtown Chicago firm may have a mainframe computer. When they set up an office in a suburb 30 or 40 miles away, they don't necessarily want to buy another big computer. Instead, they may put personal computers in the suburban office. The computers "talk" over the phone lines, through a modem, to the downtown computer. Or the company may use a dedicated leased line that will allow a company with remote sites to connect all units together and share information and peripherals such as printers, disk storage servers, and backup devices.

Why Companies Install Networks

"Even an office with three or four computers probably will be on a network," Behm says. "Sharing resources efficiently makes sense."

Using inexpensive technology, small job shop companies can cut personnel expenses, track employee data, and know the status of a customer's project—virtually instantly. One such network Behm devised and installed in a metal fabrication plant used bar coding as a management tool.

All employees wore bar-coded badges, Behm explains. They scanned their badges every time they reported for work, went on break, or worked on a customer's job. All activities for a particular job were bar coded, too.

"Management personnel can call up a software program that would tell them instantly about a particular job," Behm says. "They know which employees have worked on the project, how far the job is from completion, and whether the work has been shipped, when, and by whom."

Supporting Network Users

Behm, an independent consultant, has been helping individual users and small businesses from publishers to plastics plants solve computer problems for more than 10 years. He finds, though, that his knowledge of network technology is becoming increasingly important as more of his clients "get wired."

"My clients can only do so much with individual computer islands," he explains. "In order to connect their computers effectively, I've got to know whether dedicated wire, infrared, or other technology is most appropriate. I start by listening to the clients. 'Where are they with their computer equipment?' 'What do they want to accomplish?' Then I recommend solutions."

Behm also trains the clients—enough to solve simple problems themselves, he says, but also enough to know when to call him if complications arise. He considers himself "like a fireman who puts out big fires," on call 24 hours a day.

"It may be noon when your client finds out an ex-employee wiped out the hard drive when she left," he says, "or midnight when a lightning strike and voltage surge takes down the whole network." He has gotten a network up and running after a flood in a client's computer room, and he has restored all a customer's data after a burglary in which all computers but the file server were stolen.

NETWORK MANAGEMENT

Putting in a network is fine, but who makes sure it is functioning smoothly? In a small office, the network manager may be a secretary with a special interest in telecommunications. In today's global networks, however, advanced telecommunications and computer applications require a network manager skilled not only in technology but also in strategic business planning.

Take, for instance, a major financial institution. It will use, and possibly own, satellite transponders and earth stations, private microwave and fiber optic facilities, and sophisticated computerized switches for message and clearing functions. The institution may link its branches within a small town through dial-up access to the public-switched network, or it may have established national and international systems though dedicated packet-switched networks on private leased lines or its own facilities.

Electronic banking—and that is just one application—has created a global marketplace that requires almost instant investment decisions as well as round-the-clock access to capital; a network in which traders in Japan can take part in banking opportunities in London through a clearinghouse in San Francisco. Major international banks often operate private networks. Others lease facilities from independent vendors.

Networks are the lifeline of many businesses, both large and small. "Even a bank with 100 or 200 terminals still requires some level of network management," says Leonard Bertagnolli, director of networks and workstations for the financial systems division at Unisys.

What Is Network Management?

The term *network management* means different things to different people. IBM says it is the ability to monitor and coordinate activities centrally throughout a distributed network and provide resources that ensure the exchange of information. A top Digital Equipment Corporation (DEC) market development manager says it is the ability to manage the traffic and capacity of specific links on the network, including the ability to look down from a central site at all the components and devices connected to the network and predict their failure.

Whether network management is centralized, where an institution's control center can monitor and control everything taking place on it, or distributed, with multiple points of control at various points on the network, certain tasks need to be accomplished. The components of the network all have to work together with maximum effectiveness. This is not simple, especially if different parts of the system come from different vendors. If architectures are different, network management has to provide some kind of bridge that allows the visibility of network B within network A, if you are running both.

Another important task for network management is to identify problems and where they are occurring. With some systems, the components must be checked at predetermined intervals to make sure they are working correctly; with others, the software can spot problems with elements on the network, diagnose what is wrong, alert management, and even reroute communications around the trouble spots. In fact, some network management software even gives advance warning of potential trouble.

Security is another important problem in network management, especially when sensitive data or financial transactions are being sent over the network. Information must be backed up off-site for safety. Unauthorized use must be monitored.

Reports of network activity can and should be more than just a log of who is using the network and when. Sophisticated software can produce information on how well the network is functioning and pinpoint areas where improvement is needed. It can also show network usage, an important factor in a company's financial planning and budgeting, when tracked against departments' performance.

INDUSTRIAL NETWORKS

The same devices that communicate in an office environment—computers, modems, printers, and other peripherals—can be tied together on the factory floor. So can machines. An engineer at a company can use computer-aided design (CAD) and computer-aided engineering (CAE) to design a part, send the information from the computer by telecommunications to a machine that can be programmed to make the part, and keep all records in computer files, without ever having to print them out on paper.

In *computer-integrated manufacturing,* the term for putting all this together, data communication networks play a key role. The networks transmit data between terminals or devices that collect data to computers, between computers, or from computers to control devices. Data communication networks make it possible for companies to reduce time to market and to be flexible, thus offering significant competitive advantages.

NETWORK MANAGEMENT CAREERS

Most telecommunications experts agree that as more offices and factories automate, skillful network management will become a strategic business advantage. Today's shortage of qualified network managers makes it increasingly difficult for many organizations to keep their networks up and running. Men and women with the technical abilities to understand the sophisticated computer systems and the business ability to integrate network management with an organization's strategy will do well in this exciting field.

Even in a small, several-person office, a network manager has an important role. He or she must be able to keep the network running and handle security, making sure the company's sensitive data that is running on the network is protected from spies and saboteurs. The network manager is responsible for backup and storage, not only of files but also of software, and making sure it is done on a frequent, timely basis.

Training becomes a significant issue and is part of the hidden cost of networking. As users ask, "What can I do on the network?" "How do I use the network to help with the job I am doing?" "How will the network help me to do it faster and easier?" the network manager must know the answers—and be able to train users effectively. He or she must keep up with technology and be ready to recommend upgrading or changing the network when it is appropriate and cost-effective to do so.

A network manager must also possess "people skills"—the ability, for instance, to work closely with the office manager so the work flow is not interrupted. If the network needs to be taken down to fix one of the computers or telecommunication links, the network manager must notify the office manager ahead of time, so that the workload can be distributed effectively while a site on the network is out of commission.

CHAPTER 10

CELLULAR COMMUNICATIONS

Cellular land-mobile communications have become increasingly popular since the first commercial cellular system went on-line in October 1983. Just five years later, more than 1.6 million people in the United States were using cellular telephones and enjoying their many benefits, including increased productivity, accessibility, safety, and security. By 1988, the cellular industry had more than a $2.6 billion investment in creating its communications network.

HISTORY OF CELLULAR TECHNOLOGY

Cellular technology had its beginnings in research done by various electronic equipment companies in the late 1960s, though mobile radio had been used as early as the 1920s. At first, radio systems were one-way, primarily used by law enforcement personnel, fire departments, and other government agencies. These gradually gave way to two-way land-mobile services, which were often used for emergency operations but did not interconnect with local telephone systems.

The Federal Communications Commission has historically regulated radio communications and allocated radio frequencies, the bands within the part of the electromagnetic spectrum that are used for various forms of communication. By the mid-1940s, the FCC was allowing mobile telephone services to operate. Public paging services began operating soon afterwards. Technical developments as a result of wartime research and development made it possible for land-mobile radio and other applications to use radio frequencies substantially above the AM broadcast service band.

By 1982, the FCC had decided there should be two competing systems in each market area. One license was set aside for the local telephone company. The other license was reserved for a non-telephone company. The cellular licenses for the 305 metropolitan standard area markets in the United States were prioritized: that is, on June 7, 1982, the FCC accepted 190 applications for markets 1–30; on November 8, 1982, it accepted 353 applications for markets 31–60; and on March 8, 1983, the FCC accepted 567 applications for markets 61–90.

The FCC had granted cellular licenses for the top 30 markets after comparative hearings. By 1984, however, the FCC switched to a lottery as a way of choosing who would get licenses in the remaining markets. In order to be eligible to compete for a license, companies applying had to demonstrate legal, technical, and financial qualifications. That didn't deter applicants, however; two months after those rules had been adopted, the FCC accepted 5,182 applications for markets 91–120.

The approval procedure and lotteries continued. By December 1987, when the last lottery was held for markets 281–305, nearly 100,000 applications had been filed with the FCC for licenses. It is in these metropolitan standard areas that cellular first became operational.

HOW CELLULAR TECHNOLOGY WORKS

Motorola Inc.

Let's look at Motorola Inc., one of the world's leading manufacturers of electronic equipment, systems, and components and historically a major force in the land-mobile business. In fact, the company is the world's leading manufacturer of cellular products. Founded by Paul V. Galvin in 1928 as the Galvin Manufacturing Corporation in Chicago, its first product was a battery eliminator, which allowed consumers to operate radios directly from household current instead of having to use batteries supplied with early radio models.

In the late 1930s, when law enforcement personnel, fire departments, and other government agencies began to use mobile radio, the company successfully commercialized car radios under the brand name Motorola, a new word suggesting sound in motion. In 1947, the company changed its name to Motorola Inc., and by 1959 when Galvin died, Motorola was a leader in military, space, and commercial communications; had built its first semiconductor production facility; and was a growing force in consumer electronics. The future looked bright for Motorola and other land-mobile equipment manufacturers. But as allocated band frequencies began to fill up, users had problems. Busy signals, cross-traffic, and difficulty in getting a line were common problems.

Allotting Frequencies

The concept of frequency is one most telecommunications users don't consider. When you turn on your radio, for instance, you don't stop to think about how the signal you perceive reaches you. In fact, an overall electromagnetic spectrum extends from extremely low frequencies (a few cycles per second, or Hertz) to visible light, and up to X-rays.

The usable radio spectrum is a small portion of that spectrum. When radio waves are used to communicate information, the data, sound, or pictures sent over these waves are transmitted, picked up by an antenna, and processed by a receiver.

The spectrum can be thought of as a limited natural resource. Unlike most resources, however, it doesn't shrink with use. Once a particular transmission has been completed, the portion of the spectrum that was used during the transmission is free to be used again.

International spectrum management has become increasingly more important as developing countries have greater telecommunication needs. More than 160 nations, including the United States, are members of the International Telecommunications Union (ITU), a specialized agency of the United Nations. International spectrum negotiations involve U.S. domestic and foreign policy. U.S. participation is handled by the Department of State, the National Telecommunications and Information Administration (NTIA), and the Federal Communications Commission (FCC).

As Motorola's Jerry Orloff explains, "Frequency is like the air in a room. If only two or three people are in an office, you can breathe comfortably. When 50 or 60 people are in the same confined space, it becomes more difficult to breathe."

The FCC handles spectrum management in the private sector (private industry and state and local government). As allocated frequencies began to fill up in the late 1960s, companies like Motorola began researching new ways to solve the problems of overcrowding. The result was cellular technology.

"Cellular is another word for modules or blocks," Orloff explains. "Cellular is the name the industry agreed on, perhaps because it resembles little cells, like snowflakes hooked together."

By making the radio low powered with a number of antennas, companies like Motorola could increase by an extreme amount of usage the number of people who could use land-mobile radio. They could go from perhaps 2,000 persons under the old technology to as many as 300,000 in a metropolitan area. "And with the introduction of digital technology, the number of people able to use land-mobile radio can go as high as 1.5 million to 2 million per city," says Motorola's Jim Caile, director of marketing for Motorola cellular subscriber group.

Investment requirements for cellular were—and are—high. When companies like Motorola went from low-band frequency, which they had when the land-mobile business began, to mid-band UHF and then to a high-band VHF, there wasn't much difference in constructing radios. But cellular needed a completely different technology.

With cellular, it might cost a company from $10 to $15 million to put the system into a city before a motorist can make the first phone call. You need an extremely sophisticated technological switching system. You need computers and antennas. You need low-powered base stations that will operate from one cell to another.

Cellular technology, however, can overcome many problems of mobile telephone service. It makes it possible to offer thousands of channels, assuring a circuit for each customer when needed. Cellular systems also offer better transmission quality and privacy than the older type of mobile radio telephone service.

CELLULAR TECHNOLOGY AND USERS

An area serviced by cellular is divided into many small geographical cells. Each has a radius of about eight miles. Within each cell is a low-power radio transmitter-receiver unit that carries calls over an antenna system for as many as 72 radio channels. A sophisticated computer-controlled call switching system is centrally located in the area to control the transmitter-receiver to each cell and to perform the switching task.

As a motorist drives from one cell to another, the central system monitors his or her movements and transfers the call to the transmitter-receiver in the cell the customer is entering, without interrupting the conversation. Since the calls are carried on low-power radio signals, the same channels can be used simultaneously by different motorists in nearby cells with virtually no change or interference. Future growth needs can be met by subdividing the cells until thousands of channels can be provided within the service area.

The mobile telephones themselves are compact units that can easily be installed in a motor vehicle. There are basically three types: the regular car phone, which is installed in the vehicle; a transportable phone, which is a portable phone with a small carrying case that provides a full three watts of power; and a handheld lightweight portable phone, about the size of a home cordless phone, but with only approximately .6 watts of power.

Many cellular phones are truly eyes-free and hands-free. They are able to respond to a driver's spoken instructions, and the driver never needs to touch the mobile handset.

A user programs the cellular phone, which can be trained to recognize the driver's voice. The unit vocally stores and recalls up to 40 names for a single user or 20 names each for two users. Then the phone is voice-activated and operates on simple spoken commands to place, receive, and end calls. The driver tells the unit, "Turn on telephone." Next, the driver asks the phone to call a particular person. The unit recognizes the spoken name and dials the call to the previously entered number. When the driver has finished talking, he or she merely says, "Hang up telephone," and the unit obeys.

Some cellular phones have a special interface that works with other telecommunications devices. It is possible for a driver to plug in a facsimile machine to the cellular car phone and send a document to another fax machine anywhere in the world, all while driving on the highway. It is also possible to plug a laptop computer into the cellular phone and communicate by modem with other computers. Communications consultant David Finn says the Internal Revenue Service plans to equip their auditors with laptops and portable cellular phones, enabling them to access mainframe IRS computers from any remote location.

Calls placed over cellular systems can be linked to nationwide and international telephone networks.

RECORD-BREAKING GROWTH

The 1990s have seen record-breaking growth in most wireless service segments, including cellular, paging, and specialized mobile radio. Cellular alone had 15 million U.S. subscribers by 1993, and growth was predicted at 36 percent per year.

Cellular phone technology and the cellular market have changed considerably since 1983, when Motorola introduced the first commercial cellular mobile telephone. Today, cellular system manufacturers are developing equipment that uses digital technology. According to the North American Telecommunications Association, in 1995, 80 percent of cellular phone systems were digital; 10 percent used both analog and digital transmission, converting signals from one mode to the other; and 10 percent were analog.

The Motorola Cellular Impact survey of 1993, conducted by the Gallup Organization, found that cellular phones have had a dramatic impact on the lives of the 11 million Americans who use them, making users more successful at business, providing personal safety protection, and allowing them to make the most of their personal time. In fact, three out of four business users who responded said cellular phones increase flexibility, efficiency, and productivity and make them more competitive.

Personal safety protection was another reason cellular has become popular, according to the survey. Users said cellular phones make them feel more safe and secure and more willing to help strangers. Cellular phones allowed them to phone for roadside help for their own disabled vehicles and allowed them to phone for help for someone else's medical emergency.

CHANGES IN CELLULAR USE

The way people are using cellular telephones is changing as the technology becomes widespread. In 1991, 67 percent of cellular calls were made for business purposes; by 1993, however, cellular usage was split: half the calls were business related while half were personal.

In addition, mobile office technologies are impacting the cellular market. By 1993, 27 percent of users combined cellular telephones with a pager; 16 percent used cellular for conference calls; and 13 percent used cellular voice mail. Eleven percent combined cellular telephones and faxes, 6 percent used electronic mail, and 6 percent transmitted data from a computer over their cellular phones.

Mobile Office, a leading trade publication covering the portable computing and wireless communications industries, predicts even more explosive growth in this area. In 1994, its research found there were 27.3 million traveling professionals, managers, and business owners. Three-quarters of those who used portable computers had modems; of those, seven of ten routinely used the modems to connect back to the office when they were traveling.

PAGING

The pager industry has also seen dramatic growth. In 1992, there were 15.3 million paging subscribers. By 1995, that had grown to 19 million. Motorola predicts by the year 2000, more than 50 million Americans will be using pagers.

About one-third of U.S. subscribers use paging for non-business, and half of all new pager users are personal users, according to figures compiled by Economic and Management Consultants International, a Washington, D.C., company that follows telecommunications industry trends.

EMCI expects U.S. pager sales will grow to 40.5 million users in 1998, while the world market will grow from 40 million to 111 million users.

A SATELLITE CELLULAR NETWORK

First conceived by Motorola, Iridium, a global satellite wireless telephone system now under development, will provide communications to areas not accessible by land systems. Sixty-six low-flying satellites will be able to establish a radio link with Iridium radiophones and pagers, regardless of where the subscriber is. Earth-based gateway switching centers will link Iridium calls to existing landline and earth-based cellular systems.

Eventually, Iridium will allow persons with portable phones to make and receive calls anywhere in the world. Voice and data transmissions can be sent over mobile and portable radiophones and multiline, transportable phones. Future technologies may provide each person with a single, unique phone number—letting that person be reached anywhere at any time.

A FUTURE IN TELECOMMUNICATIONS

If you want to work in telecommunications, consider that wireless service technology is headed for spectacular growth in the coming years. Jobs will come in research and development, making and selling equipment and services, and in setting up and running networks. Reading the trades and monitoring developments may help you get one of those jobs.

CHAPTER 11

EDUCATION AND TRAINING

What should you study if you want a job in telecommunications? How and where should you get your training? There is no one right answer for everybody, because there are many opportunities and many career choices. However, here are some guidelines that may help.

Increasingly, telecommunications is becoming more oriented towards the integration of information systems. That means tomorrow's telecommunications experts will have to know about computers, data, telephone systems, connections, and image technology. In addition, top-level telecommunications positions will almost certainly go to persons who can not only set these information systems up for companies and see that they are maintained properly but who can also show senior management how to use the information from telecommunications systems to make strategic decisions.

There will always be hands-on jobs in manufacturing, for telephone operators, and for customer service representatives. But top jobs, salaries, and opportunities will require a combination of technology and management skills.

Consequently, you will want to get as much training as possible in mathematics (including calculus), science and electronics, computers, communication skills, and management studies.

DISTURBING FINDINGS

Recent studies, including a report from the Business-Higher Education Forum, an organization of 40 corporate CEOs and 40 university presidents, have shown that the United States has a critical shortage of skilled, adequately trained, and committed workers. Their report includes disturbing statistics:

- As many as 27 million Americans may be functionally illiterate.
- U.S. corporations spend over $25 billion each year to teach their workers basic skills—how to read, write, and use a computer—skills that should have been taught the first time around.
- Members of minority groups will account for one-third of the new entrants to the U.S. labor force between now and the year 2000.
- Far too few minority members are getting the college degrees and advanced graduate and professional education needed for the best jobs of tomorrow.
- At the time when America's economic survival depends on technological innovation, the United States may face a shortage of 700,000 scientists and engineers by the year 2010.

MAJORING IN TELECOMMUNICATIONS

Several college guides, available at the reference desk of your school or public library, will give you information on schools that offer telecommunications degrees or that have students majoring in telecommunications.

You can find this information in *Lovejoy's College Guide,* edited by Charles T. Straughn II and Barbarasue Lovejoy Straughn. Published by Arco, 15 Columbus Circle, New York, New York 10023, the book covers some 2,500 American colleges and universities. More than 75 schools are listed in the telecommunications section under career curricula and special programs.

A similar guide is published by the College Entrance Examination Board, 45 Columbus Avenue, New York, New York 10023. Its *Index of Majors and Graduate Degrees* lists telecommunications studies in 30 states. The *Index* also indicates whether schools offer an associate degree, a bachelor's degree, a master's degree, or a doctoral degree.

In the *Chronicle Vocational School Manual,* published by Chronicle Guidance Publications Inc., P.O. Box 1190, Moravia, New York 13118, schools are listed by categories: telecommunications electronics, telecommunications installations, telecommunications management, telecommunications technology, and telephone equipment installation and maintenance technology. Separate listings cover television broadcasting/production, television and electronics repair, and television studio operations.

Because telecommunications means different things to different people, you may find courses in this field listed under other entries. Many universities group courses in broadcast management, radio and television production, and regulatory or legal issues under the broad heading of telecommunications. Others reserve the name telecommunications for the core technical, electronics-related studies.

For instance, Rhode Island College senior Leigh-Ann Gauvin, majoring in telecommunications there, has taken courses in advanced television production, television editing, communications law and regulation, and mass communications. However, Stan Simbal, adjunct faculty member at Rockland Community College, Suffern, New York, describes the telecommunications course he teaches as "providing the necessary information to develop and integrate a communication system, using both the telephone and the computer." Simbal's students learn technology for working with connecting personal computers to mainframe hosts; Ethernet, Token Ring, and Star (three ways of transmitting data communications); and cabling requirements needed for Internet connections.

CONTACTING THE SCHOOLS

Books like *Lovejoy's* and *Barron's Index of College Majors* cross-reference schools by geographical location and offer school addresses to which you can write for further information.

When you write to a college or university, there are several questions you should be asking. You will want to know whether the school is accredited and by what agency. If it is a vocational school, it may be accredited by a state agency or belong to a national association.

Ask for a catalog and any flyers or brochures that talk about the telecommunications programs. Read the material you receive carefully. How much of the coursework seems to be theory? How much, if any, is hands-on? What laboratory facilities, opportunities for co-op programs, and help in securing intern positions can the school offer?

You will want to write a letter to the person who heads the telecommunications department or to one of the faculty members. Ask specifically how many recent graduates are working in telecommunications. What positions do they hold? What are starting salaries? Many schools with specialized departments, such as telecommunications, know their graduates well and keep track of where they are working. You can ask if they will put you in touch with an alumni whom you can contact for advice.

HOW TO FIND OUT WHAT TO STUDY

One thing is certain: telecommunications technology is changing so rapidly that you will have to keep up with developments as you plan your college courses. The convergence of technology for computers, telephones, video, broadcast television, and cable is happening so quickly that even those professionals already in the business are scrambling to monitor developments.

If you are looking at fields like network management, you will need a combination of electronics technology and management skills along with a business orientation. If you like telemarketing, you will need communication skills, along with sales and marketing training. Advertising and copywriting skills become important if you plan to be involved with interactive shopping on the Internet. Financial issues become significant in the field of electronic financial services, where telecommunications technology makes it possible to transfer billions of dollars electronically every day.

If you are interested in the educational services field, you will want to know about teaching and learning theories and about fields that interest students: science, math, and history, for examples. Writing and publishing on the World Wide Web requires computer graphics proficiency and journalism skills.

Interactive entertainment—a rapidly growing field—requires knowledge of developments in films, television, videos, music, theater, and books as well as the knowledge of how to put them on-line. For instance, Delphi Internet Services, one of the leading vendors, lets viewers download 30-second video promos highlighting season premieres of Fox television shows, photos of the show's stars, and databases containing information about the series.

One easy step to take to learn more about educational qualifications for telecommunications jobs is to read Sunday editions of newspapers from major cities: the *New York Times,* the Chicago *Tribune,* the *Los Angeles Times,* and the *Denver Post.* Look both in the help-wanted classified sections and in the larger display ad sections for telecommunications positions. Keep a file of the ads and a record of the educational qualifications required for the various positions. You can use this as a starting point for choosing what to study.

Another source of telecommunications advertising is the *Wall Street Journal* and its companion publication, the *National Business Employment Weekly.* These probably are on your library shelves and almost certainly at your neighborhood newsstand or full-service bookstore. Again, look at the educational requirements, though most of these positions are for experienced personnel.

You will also want to read the trade magazines listed in Appendix B. More than any other technique, reading the listed publications will give you the latest information on how the field of telecommunications is developing and where the growth areas seem to be.

On-line searching, described in detail in Chapter 12, is another quick way of finding out how well you measure up to what is expected of job applicants.

CD-ROM disks, available at many public or university libraries and even offered in home computer versions, allow you to search databases for companies by Standard Industrial Classification (S.I.C.) codes and will give you phone numbers and addresses. Some S.I.C. codes you can start with are telephone and telephone equipment—installation, S.I.C. 1731; telephone dialing devices, automatic—manufacturing, S.I.C. 3661; and telephone, cellular radio—manufacturing, S.I.C. 3663. You can write to the human relations department of companies you find, asking them to send you information on educational requirements for job applicants.

SELECTED SCHOOLS

To list all the schools in which telecommunications courses are offered would take a whole volume. Here are descriptions of selected programs. Contact addresses are given so you can write the schools directly for detailed information.

Arizona State University

At the Walter Cronkite School of Journalism and Telecommunication, studies emphasize mass communication and information management skills. Technical training is subordinate to academic discipline. Broadcasting and journalism majors are offered.

Broadcasting students study the theories, processes, and skills related to electronic mass communication. Their courses include work in radio, television, cable, and other emerging technologies, such as satellite communication, high-definition television, and fiber optics.

For information, write: Walter Cronkite School of Journalism and Telecommunication, Arizona State University, Tempe, Arizona 85287–1305.

Carnegie Mellon University

The first degree program of its kind in the nation, the master of science in information networking at Carnegie Mellon is also the only program that integrates the disciplines of computer science, electrical and computer engineering, and business and policy studies. The program's other unique strength is the combination of a solid technical foundation with managerial concepts essential to the strategic use of information networking.

Information networking students take core and elective courses alongside students in the colleges of computer science, engineering, and business at Carnegie Mellon. Working closely with each other, with faculty, and with students of related disciplines, information networking students experience the benefit of the many technical backgrounds at Carnegie Mellon.

The 14-month program consists of 10 months of study and a 4-month group project or thesis. Students complete core courses and elect a specialized field of study within information networking. They are given the choice to work on either a group project or thesis which gives them the opportunity to integrate what they have learned and apply their knowledge to actual industry problems. Whether they choose the project or thesis, students are mentored by information networking faculty and by business and technical experts from industry. Often, project and thesis research provides students with the seed for continuing research after graduation.

Graduates of the program manage and design technology, work in research, and may actually invent new technology.

A number of major companies sponsor their employees in the program. However, students need not have sponsors to be eligible for the program.

Write: Carnegie Mellon University, Information Networking Institute, 5000 Forbes Avenue, Pittsburgh, Pennsylvania 15217.

Columbia University

Founded in 1985 as one of the six original National Science Foundation Engineering Research Centers, the Center for Telecommunications Research (CTR) at Columbia University supports research, education, industrial collaboration, and technology transfer. All focus on integrated telecommunications networks for high-speed multimedia information movement and management.

Its programs, organized in close cooperation with 27 industrial partners, address key technological issues that confront the telecommunications community. Programs also provide a cross-disciplinary learning experience for students and produce a solid platform for technology transfer and later commercialization.

CTR is known for innovations in lightwave networks, network management and control, and broadband applications. Its ACORN project, together with 10 of its industrial partner companies, produced the world's first working prototype of an ATM-based lightwave network operating at access speeds of one gigabit/second.

Its ADVENT project has produced advances in signal processing for image and video signals. Its Student Electronic Notebook project, also with industry collaboration, has produced software and reliable access strategies for broadband mobile computing services.

CTR focuses on five cross-disciplinary projects: lightwave networks; local and wide area nomadic computing that provides services to mobile users; ADVENT, targeted at aspects of digital image, digital video/HDTV, and interactive multimedia applications; control and management of enterprises, targeted at network management and control architectures for large networks with diverse quality-of-service needs; and multimedia networking, targeted at teleconferencing the distributed work environment, including desktop publishing and picture archiving.

Other departments and schools at Columbia University are participating in projects.

For information, write: Center for Telecommunications Research at Columbia University, 801 Schapiro Research Building, New York, New York 10027.

DePaul University

Although there is no undergraduate program in telecommunications, DePaul's School of Computer Science, Telecommunications, and Information Systems offers a master of science in telecommunications systems and a Ph.D. in computer science which may be concentrated in a telecommunications area. The school also offers several professional nondegree programs through the Institute for Professional Development.

The M.S. in telecommunications systems program has around 150 students enrolled and graduates approximately 40 to 50 students per year. All classes are during evening hours, as approximately 80 percent of the program's graduate students are employed full-time. The department offers 20 graduate courses in telecommunications topics, of which 12 to 13 must be completed after prerequisites are met for the M.S. degree.

DePaul's Ph.D. program requires additional coursework and breadth examinations in other areas of computer science as well as research, publication, and dissertation.

The Institute for Professional Development offers intensive certification courses in various technologies. The Telecommunications Program is a 12-week course covering major topics in telecommunications technology, systems, and management. The Local Area Networks Program covers LAN technologies, design, and management.

For information, write: The School of Computer Science, Telecommunications, and Information Systems, DePaul University, 243 South Wabash Avenue, Chicago, Illinois 60604-2302.

George Washington University

The George Washington University's Columbian College and Graduate School of Arts and Science offers a master of arts degree program that focuses on the changes taking place within the telecommunications industry. The program is specially designed to meet the needs of the following: telecommunications managers, policy analysts, sales and marketing professionals, network design specialists, system administrators, traffic and network analysts, and communication and information specialists. The program is also suitable for those with strong academic backgrounds who desire a comprehensive education in this increasingly important field. The four areas of concentration within the program are history and policy, economics, engineering, and management.

The program consists of 12 graduate courses (36 credit hours), including 10 required courses and 2 electives selected in consultation with the program advisor. Each course carries 3 credit hours. A thesis option is available, in which case a total of 10 courses plus a 6-credit thesis are required. A Master's Comprehensive Examination is required of all candidates at the conclusion of their coursework.

The program offers students an opportunity to observe and analyze industry developments as they occur among the major companies and associations as well as state and federal agencies. Washington, D.C., is home to major offices for such important carriers as Bell Atlantic, Comsat, MCI, and US Sprint; key federal regulatory agencies, including the FCC and NTIA; and hundreds of industry associations and special-interest groups.

For more information, contact: The George Washington University, Graduate Telecommunication Program, 812 20th Street N.W., Washington, D.C. 20052.

Georgia Institute of Technology

This program aims to provide a well-balanced coverage of both communication systems and computer networking, emphasizing mastering systems-level considerations, that is, how do the various hardware and software components operate together to provide the desired services?

This program uses a series of lecture courses reinforced by several lecture-laboratory courses that allow the students to actually observe the operation of such systems, to better understand the nature of the interactions involved, and to measure their performance.

Either undergraduate or graduate students can specialize in this area; in fact, all undergraduate students in the College of Computing must take a coordinated series of courses in one or more areas of specialization.

Courses in the program are also available to students majoring in other fields, such as electrical engineering and other disciplines.

The undergraduate program provides an introduction to telecommunications systems and a complete coverage of data communications, the exchange of blocks of information between digital devices. Operating systems, programming languages, and mathematics are also covered.

Graduate students study computer networking as well as specialized topics in communication systems such as protocol design and validation, programming of control systems, and queuing theory.

Georgia Institute of Technology offers a five-year cooperative plan for students who want to combine practical experience with technical theory.

For information on the Special Program in Telecommunications and Computer Systems, write: Professor Philip H. Enslow, Jr., College of Computing, Georgia Institute of Technology, Atlanta, Georgia 30332–0280.

Michigan State University

The Department of Telecommunication offers both undergraduate and graduate degrees. The program provides students with a broad understanding of the fundamentals and emerging developments in telecommunications technology, policy, management, and economics. The department stresses the convergence between the telephone and other industries, such as cable, broadcasting, computing, and information services.

Michigan State University's program also focuses on the global nature of the industry and offers overseas study opportunities. For the more technically oriented student who desires an engineering perspective, there is a joint program with the College of Engineering.

Graduates of the programs work primarily in management and marketing positions for many different kinds of firms. Other graduates help to manage large corporate telephone, data, and video networks for such companies as Chrysler, Whirlpool, and K-Mart. Still others have taken positions as marketing and account managers for telecommunications companies such as AT&T, MCI, and Ameritech.

Some students have found positions planning and implementing networks for large telephone companies like Ameritech. Finally, many graduates go to work for consulting firms like Arthur Andersen, EDS, and Deloitte and Touche, helping other companies plan and manage their telecommunications needs.

A unique resource of Michigan State University's program is the Information Technologies and Services Laboratory, which has computer workstations, a digital PBX, full Internet access with its own World Wide Web server, and special network design and management software. Other resources include the Comm Tech Lab, a multimedia research and development laboratory, and the Telecom in Europe program, a summer course based in Paris, France, covering the telecommunications situation in the European Union.

For more information, write: The Department of Telecommunication, Michigan State University, Room 409, Communication Arts Building, East Lansing, Michigan 48824-1212.

New Mexico State University

The Department of Engineering Technology at New Mexico State University offers both associate and bachelor degree programs. Students can take courses in telecommunications, computer technology, manufacturing, electronics, mechanical technology, and civil technology. The telecommunication courses include electronic RF communications and digital and data communications, fiber-optics, satellite communications, and computer networking.

For more information, write: Department Head, Department of Engineering Technology, Box 30001 Dept. 3566, New Mexico State University, Las Cruces, New Mexico 88003.

Ohio University

Students who major in communication systems management learn about voice and data communication systems that governments, organizations, and corporations use. They study the benefits and drawbacks of communication technologies like fiber optics, satellites, computer networks, videoconferencing equipment, microwaves, and telephone systems.

The program teaches students the basics of business, management, and engineering and provides hands-on lab courses on how telephone systems operate. Also studied: legal and regulatory basics which affect the field. Juniors and seniors can compete for paid internships.

For information, write: Director, J. Warren McClure School of Communication Systems Management, Ohio University, 9 South College Street, Room 197, Athens, Ohio 45701.

Rhode Island College

The Department of Communications balances a program highlighting practical experience and theory. Flexible requirements allow the department to accommodate individual backgrounds, needs, and interests. Internships provide students with on-the-job communication experience. The department's goal: to help students gain transferable skills involving the creation, analysis, processing, and distribution of information. Graduates have found work in journalism and production at major broadcast and cable networks. Courses in telecommunications and public relations introduce and emphasize today's digitally oriented tools.

For information, write: Department of Communication, Rhode Island College, Mt. Pleasant Avenue, Providence, Rhode Island 02908.

Rochester Institute of Technology (RIT)

RIT offers undergraduate, graduate, and certificate programs in telecommunications. The undergraduate bachelor of science program in telecommunications/engineering technology has TAC/ABET accreditation and a required 15 months of cooperative work experience.

The cooperative work experience starts in the junior year. For three years, students attend classes for six months of each year and work for six months each year. Students in this program choose either the technical or management option in their junior year of the program.

Graduates of the program in telecommunications engineering technology will be able to identify network requirements for either voice or data, develop information solutions and services, understand telecommunications policy and regulation, and select the appropriate hardware and software for business environments.

Graduates have strong communications and technical writing skills and have worked in project teams.

Those graduates who have chosen the management option will also have courses in accounting, finance, organizational behavior, and marketing. The average starting salary for students graduating from the program is $30,000 a year.

RIT also offers a master of science program in telecommunications software technology and a bachelor of science degree in arts and science with telecommunications, applied computing, and management options. Both are total distance learning programs. The bachelor of science degree assumes that students have completed two years of college courses that can be transferred to RIT before starting the program. The master of science degree has bridge courses for students who have not completed the prerequisite courses for the degree.

Professional certificates in data communications, voice communications, applied computing, and telecommunications network management are also available through distance learning. Each of these certificates requires three to five courses that must be taken from RIT.

For information, write: Rochester Institute of Technology, Office of Admissions, 60 Lomb Memorial Drive, Rochester, New York 14623–5604.

University of Denver

University College is the University of Denver's college for nontraditional programs. The majority of students in their programs are working professionals attending classes in the evenings and on weekends. Courses are structured in an accelerated five-week format, meeting one night each week for those five weeks.

The master's degree in telecommunications (MTEL) is an interdisciplinary program combining studies in the technology, economics, regulation, and management of telecommunications. The program is designed to provide advanced information in the field of telecommunications both to students currently working in the industry and to students transitioning into the field. Students may custom-tailor their degree studies to reflect their individual career objectives.

The degree does not require an extensive background in mathematics or an undergraduate degree in a technical field. A baccalaureate degree from an accredited institution of higher education is required. Students entering the program should have mathematical skills in advanced algebra and trigonometry and should be familiar with statistical functions. In addition, students must have acquired a working familiarity with microcomputers.

The curriculum is comprised of nine core courses and eight electives. Core courses provide students with a foundation in voice and data communication, telecommunications policy and regulation, management, and economics. Students may tailor their electives to meet their individual personal and professional needs. A capstone project, similar to a thesis, is required for students completing the master's degree.

For more information, contact: Director, Telecommunications Program, University of Denver, 2211 South Josephine Street, Denver, Colorado 80208.

University of Miami

An innovative master's degree program offers students three choices that integrate telecommunications with business. As part of the master of business administration program, students can take a telecommunications specialization. Or, as part of the master of science degree in computer information systems, students are required to take a specific course in telecommunications introduction and fundamentals and can choose the other telecommunications courses as electives.

A third option leads to the master's certificate in telecommunications management. Students take the five primary courses in telecommunications and receive the certificate. Students who are enrolled in one of the first two options and who complete all five of the telecommunications courses receive a certificate in addition to their degree.

For information on admissions, write: Graduate Programs Office, School of Business Administration, University of Miami, Coral Gables, Florida 33124. Questions on the content or sequencing of courses should be addressed to Dr. Joel Stutz, University of Miami, P.O. Box 248818, Coral Gables, Florida 33124.

University of Missouri-Kansas City (UMKC)

Here, the Computer Science Telecommunications Program (CSTP) draws students from all over the world. CSTP faculty and researchers maintain close working relationships with scientists at companies such as Sprint Corp., Northern Telecom Inc., and Allied-Signal Corp. The program itself was started under a multimillion-dollar research agreement with United Telecom.

The Center for Telecomputing Research (CTR) works to design, develop, and apply new technologies, focusing on networked multimedia for distance education. The Open Systems Environment Laboratory (OSE Lab) tests new standards and products for the telecommunications industry and offers training in open systems as well as software development and testing.

CSTP offers two undergraduate degree options: telecommunications and computer networking, and an area of specialty in telecommunications. Students have choices of electives, such as computer networks; database management systems; computer graphics; and artificial intelligence.

The master's program emphasizes telecommunications or computer networking. Secondary concentration areas include computer architecture; computer security; data structures and algorithms; operating systems and architecture; performance modeling; and software engineering. The Ph.D. program spans several disciplines while concentrating on telecommunications or computer networking.

For information, write: Computer Science Telecommunications Program, University of Missouri-Kansas City, 5100 Rockhill Road, Kansas City, Missouri 64110.

CHAPTER 12

JOB-HUNTING TIPS

What are your chances for getting hired in telecommunications? How do you land that first job or get promoted? For that matter, what sorts of jobs exist in telecommunications?

One way to find out is to write to trade associations, asking for career literature. An excellent source of career information is the International Communications Association (ICA), 12750 Merit Drive, Suite 710, LB-89, Dallas, Texas 75251.

Your public library, and perhaps your school library, will almost certainly have on its shelves a publication from the U.S. Department of Labor, the *Occupational Outlook Handbook.* Check the index for specific telecommunications-related occupations. Listings are cross-referenced when appropriate. For each occupation, you will find information on the nature of the work, working conditions, numbers of those employed (as of the date of publication), training, other qualifications, advancement, job outlook, earnings, related occupations, and sources of additional information.

Another government publication you will find useful is the *U.S. Industrial Outlook,* which includes forecasts for selected manufacturing and service industries. It is almost certainly available at your local public library. From it, you can get an overview of important industry developments and what the U.S. Department of Commerce thinks those developments may mean in the future. Then you can consider what they may mean to you and your future in telecommunications.

For example, from the *Industrial Outlook* for 1994, we learn that potentially dramatic changes are expected in the local exchange market. The regional Bell holding companies will almost certainly strengthen their capabilities in the domestic market and will pursue business opportunities outside the boundaries of their local service area. They will attempt to develop a full range of video and information services.

That information translates to possible jobs: jobs to research and develop the services, jobs in creating them, jobs in marketing and selling them to businesses and homeowners, and jobs in accounting to track and bill for the services. One of those jobs might be yours.

You can also write to the U.S. Department of Labor, Bureau of Labor Statistics to purchase its biennially-updated bulletin *Outlook 1990–2005,* which estimates overall and sector economic growth with consistent industry and occupational employment projections. You will find, for example, that in 1990, there were approximately 325,000 telephone operators: 53,000 central office operators; 26,000 directory assistance operators; and 246,000 switchboard operators.

If government projections are correct, in 2005 there will be 205,000 telephone operators: 20,000 central office operators; 10,000 directory assistance operators; and 175,000 switchboard operators if the economy grows at a low rate. Using projections for moderate growth, in 2005, there will be 221,000 telephone operators: 22,000 central office operators; 11,000 directory assistance operators; and 189,000 switchboard operators.

At the same time, however, BLS predicts mathematical and computer scientists will be the fastest growing occupational group—with employment up 73 percent from 1990 to 2005.

What does this information mean to you?

It is important to remember that projections are only a "here's what we think" guess about the future. They are broad and general, whereas what you want is a specific job in an industry you have chosen. Nevertheless, by using this information when you explore possible career opportunities, you can be aware of events and trends that will probably influence the number of available jobs. You will be making an informed decision.

CHECK NEWSPAPER ADS

From time to time, you will find telecommunications management/ technical jobs advertised in *The Wall Street Journal* and in its *National Employment Weekly.* Both should be available at your local public library or available for purchase at your neighborhood newsstand.

Also, companies advertise telecommunications jobs in classified advertising sections of major newspapers, especially in Sunday editions.

You can study the ads to learn the qualifications required for particular jobs. One recent ad for a telecommunications coordinator wanted persons who could perform voice grade, DS1 and DS3 remote testing; retrieve, interpret, and analyze alarms; act as a liaison between customers and operations; and perform other centralized operational functions. Two to five years of telecommunications control center technical operations or related experience and/or an AA/BA/BS degree in telecommunications was required.

Another ad for technicians looked for persons who could provide support to field maintenance, performing real-time analysis of network systems and implementing hardware, software/firmware, and configuration upgrades for network elements. A technical degree, two to four years' experience in a network operations control center environment, and knowledge of telecommunications technology were required. Knowledge of radio systems, T1, multiplexers, and X.25 packet switching networks was desired.

On the same day, a telecommunications placement agency was looking for design and software engineers, marketing managers, and account managers. Another wanted a senior software network engineer. Hewlett-Packard was recruiting for six permanent positions in wireless communications, and U.S. Robotics had 11 positions open in telecommunications-related jobs.

Sunday newspapers' classified sections will often carry ads for telemarketing jobs. Pay close attention to the wording of the ad, which usually indicates whether you will be accepting calls (inbound telemarketing) or required to cold-call and canvass for leads (outbound telemarketing).

An ad reading, "Make $6 to $10 an hour salary, setting appointments for home improvement company, flexible hours, must have experience in telemarketing remodeling, carpet, vacation plans, and books," is an indication you will probably be required to pitch the clients' products, often through randomly dialed calls. Phrases like "real go-getter," "self-motivated starter," and "sales hot shots," along with a salary range listed in the ads, "earn $220-$450/week salary plus commission plus bonus," could indicate that your job security depends on quick, profitable results.

On the other hand, an ad for telephone interviewers conducting surveys that offers, "No selling, no solicitation. Interesting, challenging assignment for people seeking part-time employment 25-35 hours a week. No experience required. Good pronunciation and dictation a must," may be a less-pressured way for you to try out telephone work.

TRADE PUBLICATIONS

Another way to get some idea of telecommunications jobs available is to look at the reader service information cards bound into some of the trade magazines. Subscribers seeking more information about advertisers are asked to describe their job titles in categories:

- inside/outside plant technician, specialist, installer, analyst
- sales, marketing, customer service representative, account executive, business analyst
- support staff, including legal, regulatory, public relations, finance, MIS/DP (management information systems/data processing)

Organizations, too, are listed by categories: interconnect/integrated systems providers (Ameritech Communications); local exchange carriers (Bell operating companies, GTE, United, Contel); long-distance or interexchange carriers (AT&T Communications, MCI, US Sprint, ALC, NTN); regional or independent holding companies that don't fit the other categories; and business or corporate telecom users.

Equipment also is listed by category. Lists include central office switching equipment, transmission, outside plant, customer premise equipment, and inside and outside network support equipment. If you think you might like to work for a vendor or distributor of this equipment, you can look up any of the categories mentioned in *Thomas Register of American Manufacturers and Thomas Register Catalog File,* a multivolume set sure to be on library reference shelves. You can find out just who makes the equipment and where the company is located. Then contact the company directly to learn about their hiring practices.

Trade magazines also run help wanted ads. Though most of these are for experienced personnel, some ads list recruitment firms along with phone numbers and addresses.

THE FUTURE FOR TELECOMMUNICATIONS JOBS

To those who track the industry, changing technology, higher telecommunications costs, and the growing critical nature of information processing are pluses for a long-term increase in telecommunications jobs. Countering that, however, are consolidations, which have eliminated thousands of positions, many of them previously held by older, telephone-industry middle managers.

In addition, employment at telecommunications equipment manufacturers is down. The reasons are complex and not altogether clear. However, the high value of the U.S. dollar compared to foreign currencies has resulted in an increase of imported merchandise into the United States. It has been hard for domestic-based firms to compete with cheaper imports, and older, less-efficient plants have been phased out. In some cases, the design and engineering of products is done within the United States, but much of the actual assembly is done overseas. Products come into the United States under the brand name of the domestic parent.

IMPROVING YOUR CHANCES

There are a number of things you can do to make it easier to land that first telecommunications job. One of the most basic is to get the best education you possibly can. That means doing your top work in school and getting a thorough grounding in math and science. The U.S. Department of Education's National Center for Educational Statistics says that only 5 percent of all twelfth-grade Americans tested in 1990 were proficient in reasoning and problem-solving involving geometry, algebra, and beginning statistics and probability. Yet many telecommunications jobs require individuals with those same math competencies. You will improve your chances of getting them if you are adequately prepared. Calculus and physics are often required for you to qualify for jobs requiring a technical background. So are oral and written communications skills. Because the global telecommunications market is growing, knowing a second—or even third—language may give you an advantage when you look for employment.

Trade Publications

Technology is changing so rapidly that you will have to take an aggressive role in learning about industry developments. That means keeping up with trade publications, even though you may still be in high school or college. Your local public library or nearby college library will be able to help you locate the magazines and books listed in Appendix B.

One effective strategy is to subscribe to one or two of the leading trade magazines. Although the subscription or purchase cost may seem like a lot of money, especially if you are still going to school and not employed full-time, the knowledge you will get from reading a top trade publication will give you an edge over your contemporaries.

Conferences

Read the trades to learn when and where major conferences are scheduled. Plan to attend if you can. Two of the giants are the ICA conference (write International Communications Association, 12750 Merit Drive–Suite 710, LB-89, Dallas, Texas 75251-1240, for information) and InternetWorld, held in fall and spring (write Mecklermedia, 20 Ketchum Street, Westport, Connecticut 06880). You will see the exhibits, learn about new technologies, and have the opportunity to talk with telecommunications professionals. Also, companies with open positions will often recruit at the conferences; sometimes they set up appointments with applicants in advance. Collect business cards of those you meet, and give them one of yours. Then follow up with a letter, fax, or e-mail, reminding them of you and asking about job openings.

USING THE INTERNET TO FIND JOBS

Looking for jobs electronically and posting resumes on the Internet are two good ways in which telecommunications students can expand their job search. Several newsgroups have information abort jobs, among them, **misc.jobs.offered**, **misc.jobs**, **misc.jobs.contract**, and **u.s.jobs.misc**. Not all of these are telecommunications related, of course, but many are.

Jobs generally are coded, and it is important to use the appropriate code when you contact a company. Most of them want resumes and applications to come in by e-mail. Job descriptions are short but to the point. For instance, one recent listing described UNIX development/testing-fileservers, Sunnyvale, CA, and asked for a reply to an e-mail address. Sometimes company names are given, sometimes not. Jobs can be technical (interactive cable television software developers) or managerial (electronic systems manager).

Some employment agencies—including those who handle telecommunications-related companies—have gone on-line. The Interactive Employment Network offered by E-SPAN links applicants, recruiters, and human resource professionals in Fortune 500 companies as well as developmental and technical organizations.

E-SPAN is searchable by keyword, region, industry, date posted, and job title. Or you can post your resume to the IEN database at its e-mail address: **http://www.espan.com**.

For more information, write E-SPAN, 8440 Woodfield Crossing Boulevard, Suite 170, Indianapolis, Indiana 46240.

Often, telecommunications-related companies will post job openings directly on the Internet, describing their organizations and positions. Novell—the network giant, with 40 million NetWare users on 2 million servers—says it is refining the way most of us think about communication. "We're connecting people with people and the information they need, enabling them to access it anytime, any place." Novell's invitation: "Join us as we continue to develop this new network of information, soon to be an integral part of everyone's life."

Typical jobs open have included technical support engineers for key/ netwire/ and TDT; electronic information engineers; or Group-Ware personnel. Applicants can fax or e-mail their resume to the company.

"Explore the Employment Opportunities," suggests Microsoft Corporation. Using its Web site (**http://www.microsoft.com**), you can learn about positions in a number of areas. Among them: Networking and Operating Systems, International; Microsoft Consulting Services; and Advanced Technology.

Bill Gates, chairman of the board and CEO of Microsoft, puts it this way, "Microsoft's future is all about providing access to the new world of thinking and communicating. The possibilities are limitless, and so are the challenges. That's why there's never been a more exciting time to be at Microsoft.

"We're looking for bright, curious, enthusiastic people who want to create the next generation of products that enable people to take advantage of the awesome power of personal computing."

Microsoft offers special opportunities to new college graduates (within two years of graduation) across a wide range of technologies as well as internships. Job codes differ for new graduate opportunities and for intern opportunities; you will need to use the appropriate one when you reference the listing.

A typical posting for a job open to new graduates describes responsibilities and qualifications. For instance, those who applied for the position of software design engineer were told they could be part of a technical team developing software for personal computer systems and applications. Their responsibilities would require working with one or more program managers, other SDEs, and test engineers to define product specifications, solidify a schedule, and design and write the code for products. Projects might include personal and business applications, networking software, multimedia, operating systems, graphical user interfaces, and integrated development environments.

Qualifications include proficiency in C or C++. Microsoft prefers academic or industry experience in developing applications, systems, compilers, or other development tools. Applicants should have been pursuing a bachelor's or graduate degree in computer science or a related technical discipline.

Another posting—this one for support engineers—called for persons to work directly with a wide variety of Microsoft customers to provide technical solutions via telephone quickly and reliably. Support engineers assist Microsoft's customers to optimize the use of their Microsoft applications and operating systems, including Windows, MS Office, Fox-Pro, Publisher, and the variety of products available from Microsoft HOME.

Qualifications for this position included an understanding of microcomputer platforms, environments, and applications. Microcomputer programming experience was strongly desired for positions available in Operating Systems End User Support. Applicants should have been pursuing a bachelor's or graduate degree in computer science, electrical engineering, MIS, mathematics, or a related technical discipline. Applicants should also have possessed excellent communications skills and a strong desire to work with Microsoft's customers to help them maximize their investment in Microsoft products.

Microsoft lists detailed instructions on how to apply for these and other positions. The company also has a 24-hour jobline with an expanded listing of opportunities. The jobline number is toll free: 1-800-892-3181 or 1-206-936-5500.

NETWORKING

Use networking to find jobs. Many times you meet others in the industry at area luncheons or workshops. You work with them on committees or organizations. They get to know your strengths and your background. When someone in their company gets promoted and there is an opening, they may recommend you.

Networking groups of small business owners or across-industry groups often list meetings and speakers in local newspapers. Even if you are not a member, you can usually pay a small fee and attend as a guest. The contacts you make are a start, and the knowledge you gain makes it worthwhile to go.

Getting that first job in telecommunications may actually be easier than it is in some other fields. Work-study programs, summer internships, and the willingness to take whatever position you can find, gaining experience while growing professionally through study and training will surely help. But your active role in monitoring the industry, keeping up with developments and technology, and your desire to play a part in tomorrow's exciting world of telecommunications may be the key for you.

CHAPTER 13

WOMEN AND MINORITIES

There is good news and bad news about opportunities for women and minorities who want to work in telecommunications. The good news is that sex and ethnic origin are no barrier to qualified candidates. In other words, if you have the prerequisite skills and desire, you have excellent chances of finding employment. The bad news is that proportionately fewer women, African Americans, and Hispanics are enrolling in engineering programs or becoming technically proficient.

Although it is certainly not necessary to be a licensed engineer to work in telecommunications, the qualifications necessary to succeed on the technical and management levels in corporations are becoming increasingly more complex. And keeping up with technology requires understanding the underlying concepts.

You had better be good enough to understand some basic electronics and to read technical specifications, installation manuals, and user guides for products. You should also be savvy enough to know what is smart, and what isn't, for your company. And this is just for starters.

WOMEN IN PHYSICS

At the heart of many telecommunications applications is an understanding of physics. Statistics on women in physics from 20 different countries on 6 continents show that the overall status of women in physics is low. Researchers looked at the percentage of physics bachelor's and Ph.D. degrees granted to women and the ratio of women to men among tenured physics faculty and Ph.D.-level research staff to obtain their ratings. Using standards they developed through this method, they determined that the United States ranks only in the lower third, along with Australia, Denmark, Japan, and the United Kingdom.

One important reason more women do not excel in science is their perception of how much time they will spend in the labor force. If a young woman expects to spend her major efforts in child rearing, she has little incentive to choose career options that require substantial educational commitment. But many young women do not realize that substantial numbers of higher paying jobs require a more quantitative educational background. They pass up science and math courses early in life, thus closing the doors on many enjoyable and financially rewarding jobs they might like to hold.

THE AAUW STUDY

A 1992 landmark study sponsored by the American Association of University Women, "How Schools Shortchange Girls," brought together results from 35 previous reports. The AAUW study concludes that girls are not receiving the same quality, or even quantity, of education as boys. Says the study: "A well-educated work force is essential to the country's economic development. Yet girls are systematically discouraged from courses of study essential to their future employability and economic well-being. Girls are being steered away from the very courses required for the productive participation in the future of America, and we as a nation are losing more than one-half of our human potential.

"By the turn of the century, two out of three new entrants into the work force will be women and minorities. This work force will have fewer and fewer decently paid openings for the unskilled. It will require strength in science, mathematics, and technology—subjects girls are still being told are not suitable for them."

The AAUW study found differences in how teachers interact with students. One researcher has reported that when teachers need help in carrying out a demonstration in science classes, 79 percent of the demonstrations were carried out by boys. Another study says that by third grade, 51 percent of boys and 37 percent of girls had used microscopes, while by eleventh grade, 49 percent of males and 17 percent of females had used an electricity meter. A third study found that girls saw the perceived competitiveness of engineering as a major barrier to women's entering the field.

EARLY CHOICES

Based on a Rockefeller Foundation study by Sue E. Berryman entitled "Who Will Do Science?" the Committee on the Status of Women of the American Physical Society has reported some disturbing conclusions. "By ninth grade," their newsletter said, "over one-third of those who will later earn a quantitative bachelor's degree already expect to pursue a career in science. By the end of twelfth grade, the pool is fully established. A necessary component of the pool's educational profile is the completion of advanced high school math. This optional sequence, which is elected by one-third fewer girls than boys, marks the first point at which the educational profiles of the two sexes diverge."

Females, then, are skipping courses that could lead them to well-paying and challenging science careers. In particular, they are turning off early on the math they need to succeed.

TWO CRITICAL DECISION POINTS FOR WOMEN

The Committee on the Status of Women in Physics (CSWP) publishes a quarterly newsletter. It has reported two major decision points for women: points where choices must be made about future educational investment.

The first point comes much earlier than most students realize. By ninth grade, over one-third of those who will later earn a bachelor's degree in quantitative fields (math, physical or biological sciences, computer science, engineering, and economics) already expect to be scientists. By the end of twelfth grade, the pool is fully established, that is, essentially all those who will go on to a bachelor's degree in quantitative science have already decided to do so.

To succeed in any of these technical fields, you need a strong math background. Yet advanced high school math is elected by one-third fewer females than males.

Of those who completed the necessary high school math sequence, according to the Scientific Manpower Commission, only 21 percent of the young women, as compared to 51 percent of the young men, chose a quantitative field when declaring their undergraduate majors.

"These two turning points account for more than two-thirds of the extra loss of women compared to their male peers all the way through the Ph.D. degree," says the CSWP *Gazette.*

MINORITY RECRUITMENT

Many engineering schools and companies, including those in telecommunications, recruit hard for qualified minorities. Most engineering schools have special programs for culturally disadvantaged students. Minority students often take advantage of these programs, which include summer programs before the freshman year; developmental year programs, which have special courses; and tutoring.

Notable is Georgia Tech's freshman engineering workshop, which invites ninth-grade students from minority backgrounds (including African American, Hispanic, and Native American) to visit the campus daily, meet with faculty, visit companies that employ engineers, and plan careers in engineering or related fields. More information on this program is available from the Director of Special Programs, College of Engineering, Georgia Institute of Technology, Atlanta, Georgia 30332.

Similar enrichment, awareness, or recruitment programs often exist at universities in a metropolitan area. Illinois Institute of Technology, for example, offers programs for talented minority Chicago-area youngsters. Ask your school guidance counselor for details, or write to an affirmative action officer at a college or university near you.

In Colorado, the industrial advisory board for the Colorado Minority Engineering Association works hard to get more minority young men and women interested in math, engineering, and science opportunities. Its program goes from seventh grade through high school graduation. To join, students have to take a science core curriculum. "Otherwise," as one telecommunications professional puts it, "if they decide in their junior year in high school that they want to be an engineer and they haven't taken algebra, it's essentially too late."

Two associations you will want to check out, if appropriate, are the National Society of Black Engineers, 1454 Duke Street, Alexandria, Virginia 22313-5588; and the Society of Hispanic Professional Engineers, 5400 East Olympic Boulevard, Suite 306, Los Angeles, California 90022. Each seeks to increase the number of minority graduates in engineering and technology. Both associations sponsor seminars and workshops geared toward preparing students for careers, the industry, and leadership roles.

SOCIETY OF WOMEN ENGINEERS

A major organization that tracks the achievements and statistics of women in engineering is the Society of Women Engineers (SWE). SWE offers pamphlets and information on scholarship programs, convention news, an SWE information packet, and subscription information for *U.S. Women Engineer,* a magazine women will find invaluable for latest findings.

Send a self-addressed, stamped envelope to SWE at its headquarters at 120 Wall Street, 11th Floor, New York, New York 10005. Ask for "FACTS: An Introduction to the Society of Women Engineers." Returning the coupon found in FACTS will get you on the mailing list for additional information. Bulletins and application forms describing the 38 scholarships SWE administers are also available.

Deadline dates are extremely important. Applicants for freshmen and re-entry scholarships can receive applications from March through June. Those who apply for sophomore, junior, and senior scholarships can get applications only from October through January.

AT&T SPECIAL PROGRAMS

"To be a world leader in providing systems and services for advanced information management and communications technologies," says AT&T, "America needs highly trained, dedicated young scientists and engineers skilled in advanced technologies."

The AT&T engineering scholarship program is designed to increase this talent pool by assisting outstanding minority and women high school seniors who have been admitted to full-time computer science, computer engineering, electrical engineering, mechanical engineering, or systems engineering studies at accredited four-year colleges or universities.

AT&T defines *minority* as underrepresented minorities and females: African Americans, Hispanics, and Native Americans. Winners, who must be U.S. citizens or permanent residents, are chosen on the basis of scholastic aptitude, academic performance, rank in class, strength of high school curriculum, and clear demonstration of motivation and ability. AT&T awards 15 scholarships each year and receives 700 to 800 applications annually. Applications and supporting documents must be received no later than January 15 for the following school year. For information on requirements, write ESP Administrator, AT&T Bell Laboratories, Crawfords Corner Road, Room 1B222, Holmdel, New Jersey 07733. Resumes may be submitted electronically; write to the appropriate address for details.

AT&T's programs include:

- the *Summer Research Program for Minorities and Women.* Qualified African Americans, Hispanics, Native Americans, and women
have the opportunity for technical employment experience at research and development labs in New Jersey and Pennsylvania. The
program is primarily directed towards undergraduate students who
have completed their third year of college.

    Recipients are chosen on the basis of academic achievement,
personal motivation, and compatibility of student interests with
current AT&T Bell Laboratories activities. They have the chance to
work at Bell Labs for at least 10 weeks starting in early June and
are paid salaries commensurate with those of regular AT&T Bell
Laboratories employees who have a comparable education. Each
summer employee is reimbursed for air or surface travel expenses
to and from New Jersey.

    Applications and supporting documents for the Summer Re-
search Program must be received by December 1. Students can get
information and application forms from Special Programs Manager—SPR, Room 1B222, AT&T Bell Laboratories, Crawfords
Corner Road, Holmdel, New Jersey 07733-1988.

- *Graduate Research Program for Women.* Two types of financial
awards are offered: fellowships and grants. Women take part in the
program beginning at the onset of their graduate education. The
program provides financial support for outstanding women who are
pursuing full-time doctoral studies in chemistry, chemical engineering, communications science, computer science/engineering, electrical engineering, information science, materials science,
mathematics, mechanical engineering, operations research, physics,
and statistics. They also have the chance to take part in ongoing research activities at AT&T Bell Laboratories. Four fellowships and
six grants are currently awarded annually in early April to women
beginning doctoral studies the following September.

    Applications must be received by January 15, with all support-
ing material in by January 31. Forms and information are available
from Dr. David W. McCall, Director, Chemical Research Labora-

tory, c/o Administrator, GRPW, Room 3D304, AT&T Bell Laboratories, 600 Mountain Avenue, Murray Hill, New Jersey 07974.

- *Cooperative Research Fellowship Program* for outstanding beginning graduate students who are members of underrepresented minority groups. The fellowship is for graduate work leading to the doctoral degree in chemistry, chemical engineering, communications science, computer science/engineering, electrical engineering, information science, materials science, mathematics, mechanical engineering, operations research, physics, and statistics. Since the program began in 1972, 124 fellowships have been awarded.

  Applications and supporting information for fellowships must be received by January 15. All supporting material must be in by January 31. Application forms and information can be obtained from Paul A. Fleury, Director, Physical Research Laboratories, c/o Administrator, CRFP, Room 3D304, AT&T Bell Laboratories, 600 Mountain Avenue, Murray Hill, New Jersey 07974.

- *AT&T Dual Scholarship Program* encourages outstanding minorities and women to pursue full-time college studies at the Atlanta University Center.

  Under the dual degree program, which provides full tuition and mandatory fees, textbooks, room and board during the school year, 10 consecutive weeks of summer employment, guidance of a mentor, and other benefits, students earn two degrees in five years: a B.S. in mathematics or physics from Morris Brown, Clark, Spellman, or Morehouse colleges; and then a B.S. degree in computer science, electrical engineering, or mechanical engineering from the Georgia Institute of Technology, Rochester Institute of Technology, Boston University, or Auburn University.

  Each year AT&T awards three new scholarships and maintains 15 students in the program.

  Applications and supporting documents must be received no later than December 31. For information and application forms, write to the Dual Degree Administrator, AT&T Bell Laboratories,

Room 3D304, 600 Mountain Avenue, Murray Hill, New Jersey 07974.

SUMMING IT UP

The word from those working in all phases of telecommunications is that sex or ethnic background is not a barrier to hiring. In fact, given the increasingly global telecommunications field, the ability to speak a second language and to understand another culture may indeed be an asset. Women and minorities who are qualified and who want to succeed certainly can. Companies are actively recruiting candidates with skills and potential. Today, your opportunities in telecommunications are limited only by your desire, willingness to learn, and your job performance.

CHAPTER 14

INTERNATIONAL OPPORTUNITIES

Telecommunications and related subjects will play a major role in Canada's future—and in fact, already have been responsible for economic expansion. The Information Technology Association of Canada says that nearly half of all Canadian jobs involve the use of information technology.

In Europe, telecommunications has been a popular field of study in universities—perhaps because of the International Telecommunication Union's studies that indicate the industry is relatively recession-proof. And international studies carried out by ITU recognize that there is a strong relationship between an effective telecommunications infrastructure and economic development. There are many career opportunities in countries other than the United States for those working in telecommunications.

TELECOMMUNICATIONS IN CANADA

Canada has been part of telecommunications even before the word was coined. After all, it was Alexander Graham Bell, a Scottish immigrant with an interest in speech and hearing, who first developed the telephone—in Brantford, Ontario.

Ever since those early days, Canada's and Canadian companies' role in internal telecommunications has been strong, primarily because of the country's geography. Nearly eight of every ten Canadians live within 200 kilometers (120 miles) of the Canada-United States border. Beyond this corridor lie rural areas where there are four people per square kilometer (.4 square miles). Still farther lie remote areas—more than 80 percent of Canada's land mass—where only 1 percent of Canadians live. Consequently, Canada has developed advanced communications systems as it plans for the future.

With revenues of $43 billion, Canada's information technology sector represents 6 percent of Canada's gross domestic product, employs 300,000 people, exports $10 billion annually, and spends $2.1 billion on research and development investment. Telecommunications represents a big part of that sector. In fact, a government study in 1992 found that 34 of Canada's top industrial research and development spenders are telecommunications companies.

Several agencies are heavily involved in tracking information technology and in programs to promote its effective use by Canadians.

Information Technology Association of Canada (ITAC). The   Information Technology Association of Canada, which represents more than 450 companies, considers itself the voice of the information technology sector in Canada. Its members include small, medium, and large companies in computer and telecommunications hardware, software, and services, including electronic information providers.

ITAC sponsors *Softworld,* the annual alliance-building software industry event in Canada, a showcase for Canadian industry. A program called Support and Promotion of Information Retrieval through Information Technology (SPIRIT) promotes the interests and profile of the content sector while the Supplier Development Program joint venture with Industry Canada connects the Canadian information technology sector with the world market.

ITAC's affiliates include the British Columbia Technology Industries Association, the Software Industry Association of Nova Scotia, and ITAC Ontario. Information on ITAC is available from 2800 Skymark Avenue, Suite 402, Missassuga, Ontario L4W 5A6.

Telecommunications Research Institute of Ontario (TRIO). Ontario has been a leader in telecommunications research, since 24 of Canada's major research and development spenders are in that province. Recent plans have included the Regenerative Digital Satellite project, which integrates digital beamforming antenna with satellite channel coding and on-board processing research. Although funding cutbacks have somewhat curtailed TRIO's planned program of university-industry partnerships in shared research, the organization continues its mission to respond to industry's perceived needs for future research personnel.

Ottawa Carleton Research Institute, a similar organization, hosts a Valley Technology Showcase yearly, featuring high-technology exhibits from regional manufacturers.

For information on both, contact Ottawa Carleton Research Institute, 340 March Road, Suite 400, Kanata, Ontario K2K 2E4.

The Beacon Initiative

The Beacon Initiative is one of the most ambitious national contributions to the Information Highway announced anywhere in the world. It includes several interrelated undertakings and investments by the Stentor Alliance, made up of Canada's full-service telecommunications companies. These investments are creating the world's longest, fully digital, fiber-optic network. The Information Highway that Canada is planning and implementing integrates present cable, telecom, broadcast, wireless, and satellite networks.

A major research study, "The Economic Potential of the Information Highway," examines the promise of Canada's new technological frontier. Authors and economists Dale Orr and Ron Hirshhorn have considered regulatory, financial, and public policy issues involved in setting up the networks.

The Stentor Alliance

Eleven major Canadian telephone companies have joined together to pool resources and to share their collective expertise in engineering, research, product development, and marketing. Together, they own three Stentor companies: Stentor Resource Center Inc. (an engineering and marketing company); Stentor Telecom Policy Inc. (the government relations advisory and advocacy arm for Stentor partners); and Stentor Canadian Network Management, which coordinates the operation and maintenance of Canada's national public telecommunications network.

Companies that make up the Stentor Alliance employ 100,000 Canadians. They generate annual operating revenues of more than $13 billion (Canadian dollars) from such innovative products as desktop conferencing. Using state-of-the-art technology, users can communicate with others in a remote location, combining voice, image, data, text, and graphics with screen-sharing. Users and their conference partners can see the same data on each others' PCs and can modify data on the spot.

Stentor companies also invest $4 billion in capital expenditures, spend more than $8 billion annually in operating expenses, and have invested more than $40 billion in Canada's nationwide telecommunications system.

SchoolNet: Canada's Educational Networking Initiative

One of the projects the Stentor Alliance has backed is SchoolNet, supported primarily by the Canadian government at federal, provincial, and territorial levels, with additional funding from several private companies.

Schools taking part in SchoolNet have access to the Internet through a variety of networks and providers: CLN in British Columbia; SaskNet in Saskatchewan; and NEWBED in New Brunswick. In Alberta, schools reach the Internet through universities and the Alberta Research Network; in Ottawa and Victoria, they access the Internet through free-nets. SchoolNet can also be found on the National Capital Free-Net in the Science and Engineering Technology Centre.

Students and teachers using SchoolNet can tap into resources far greater than they could obtain in their home communities, including 42 libraries and the National Library of Canada. In addition, over 400 scientists, engineers, and other advisors from around the world provide on-line expert advice.

For students who would like to learn more, a Career Selection Guide, available on-line through SchoolNet, lists educational requirements, expected salaries, and projected demand for chosen fields of work. Contact ITAC at 2800 Skymark Avenue, Suite 402, Missassuga, Ontario L4W 5A6, for information on how to access this guide.

OPPORTUNITIES IN THE EUROPEAN UNION

French student Thomas Douet typifies the European academic interest in telecommunications. Douet, who studies at the Ecole Nationale Superieure des Telecommunications in Paris, has concentrated on signal and image processing and on networks. Courses he has taken include math and electronics, computer science and languages (Prologue, LISP), algorithmics, databases, systems, signal processing methods and shape recognition, and artificial intelligence. In recognition of the global aspects of telecommunications, Douet also has studied English, Spanish, German, and Japanese.

Global information networks have been the focus of the European Commission, which has been dealing with the regulatory, infrastructure, and societal aspects of the information era.

In 1995, leaders of seven leading industrialized nations (Canada, Federal Republic of Germany, France, Italy, Japan, the United Kingdom, and the United States), collectively referred to as G-7, met in a conference to discuss the Information Society.

The ministers who took part in the meeting found common ground on many framework issues, including the liberalization of telecommunications, the protection of intellectual property, and data security. They also endorsed fair competition, universal access, and international cooperation as vital to a global information infrastructure. Countries in transition and developing countries must be provided with the chance to fully participate, they said. The ministers believe that the global information society will open opportunities for such countries to leapfrog stages of technology development and to stimulate social and economic development.

Such objectives are not new. Even before the Maastrict Treaty had been drafted, the European Union had realized the importance of a trans-European network for telecommunications. By the time the Commission's White Paper on *Growth, Competitiveness and Employment* was published in 1993, the information revolution had begun.

The G-7 ministers have taken the position that any new global Information Society must be built on a set of common rules, a tolerance of diversity, and habits of collaboration. Doing everything to ensure a smooth and effective transition to such a society is one initiative which all experts say will be amply rewarded with more jobs.

To demonstrate their commitment, they have launched 11 pilot projects. These range from electronic links to libraries and museums to a global emergency management system and to crosscultural training and education.

Information about the European Union is easy to obtain electronically. The European Commission has opened two servers on the World Wide Web: EUROPA and I'M_EUROPE. Users can easily move between them. EUROPA is designed to provide up-to-date information on the objectives, institutions, and policies of the European Union, including press releases and news; information on the Commission itself and its policies; and agendas of main events and calendars. I'M_EUROPE is aimed at promoting the European electronic information market. It includes material describing the functions of the Directorate General XIII for Telecommunications, Information Market, and Exploitation of Research, along with contact information. Also included on I'M_EUROPE: programs related to the information market, including their duration, status, funding, contracts, contact persons, and *Official Journal* references.

In addition, it is possible to use a general Internet gopher to connect with two database hosts: ECHO and EUROBASES.

Most databases on ECHO are free (except for telecommunications charges to connect to them). You will find several of them especially useful. Information Market Forum (FORU) lists consultants and companies offering services or partnerships related to EU electronic/information market programs. *I&T Magazine* offers the electronic full-text version of the magazine on EU telecommunications policy and developments. NEWS ON provides news on European information services and market developments and on ECHO databases. EMIR is a glossary of European employment and the labor market.

Nine additional databases offer information on the EU's research, technology, and development programs.

EUROBASES is an on-line access to the principal bibliographic, legal, and statistical databases offered by the European Commission. However, there is an annual subscription fee and time-use charges that vary, depending on the base.

For information on these services and how to reach them via Internet, write to Office of Press and Public Affairs Delegation of the European Commission, 2300 M Street NW, Washington, D.C. 20037.

FINDING JOBS INTERNATIONALLY

One of the most useful sources for finding jobs in telecommunications is Papyrus Media's Careers On-Line, a database of international employment opportunities. Available on the Internet on the World Wide Web at **http://www.ideaf.com/vendor/jobs/main. html** are lists of network and communications jobs offered by a variety of participating agencies. Listings are short, with few details; typically, they include the job location by country and region, the job specifics, such as Strong Protocols Routers or GUI UNIX Server Novell/NT Server, System Design. If you are not sure what the labels mean, you are not ready to apply. Job listings also mention the contracting agency doing the recruiting. If you want to know more about the job or wish to apply, there is an on-screen application form, keyed to job reference numbers.

GLOSSARY

ACD Automatic call distributor. A system that allocates incoming calls evenly to those answering phones.

Alphanumeric A combination of letters and numerals.

Amplification A process that enlarges an input. Usually voltage, current, or power (or a combination of these) is amplified.

Analog A signal which is a continuous variable. The original telephone system was designed to handle analog signals. In traditional voice phone conversations, the electrical currents in a telephone wire are analogs of the continuously variable sounds that the human voice produces.

ANSI American National Standards Institute.

Archie A tool to help people find things on the Internet. Archie servers maintain lists of files available for file transfer protocol (FTP), an archiving system that lets users find programs, data, or text files that are publicly available on the Internet.

ASCII American National Standard Code for Information Interchange, a standard file format that lets different types of computers interpret information in the same way.

Baud rate The rate at which information is transmitted between two devices. When the baud rate is lower, the rate of transmission is slower. When a computer user connects to an information service, there may be different phone numbers and charges for each baud rate the service supports.

Bits See **Data bits.**

Bits per second The number of bits sent for each character represented.

Bridge   A way of connecting one circuit in parallel with another. In a local area network (LAN), a bridge links two networks. The networks can be separate, or they can operate differently. Some bridges are external hardware. Others are adapters inserted into the file servers of the local area networks you want to connect.

Broadband   A bandwidth greater than twenty kHz that is wide enough to carry several voice channels.

Bus   In networks, a bus is one of the three basic network topologies, which is the way in which the communicating points are arranged. The bus topology uses one contiguous piece of cable, called the trunk. Cables, or drops, are attached through taps to connect the stations. In a bus network, messages transmitted are received by each station in the network. That is because each device is connected to a single line. Other common network topologies are star and ring.

Byte   The amount of memory required to store one character on a computer. One kilobyte, or 1K, equals 1,024 bytes.

Call Detail Recording (CDR)   An AT&T name for its proprietary cost accounting system, allowing messages to be tracked and their cost allocated to the proper user.

CCITT   See also **International Telegraph and Telephone Committee.** CCITT is a permanent study group within the International Telecommunication Union, a specialized agency of the United Nations. The European members of the CCITT are the individual country telecommunication administrations. The United States member of CCITT is the U.S. Department of State.

Centrex   A type of telephone service for business. Users of a Centrex system can dial directly outward from their extensions. A person calling from the outside can dial the desired extension directly.

CEPT   The acronym for the European Conference of Postal and Telecommunications Administrations.

CO   Central office.

Codec   A contraction of coder-decoder. Used especially for video conferencing, a codec device converts analog signals into digital form for transmission and converts them back again at the destination. The device removes redundant information during the encoding process.

Data bits   The individual component binary digits (bits) used to transfer data from one source to another. These bits are a series of 0's and 1's.

DDD   Direct distance dialing. A person using DDD does not need an operator to reach telephones outside the user's area. He or she can dial them directly.

DECnet   A proprietary Ethernet network from Digital Equipment Corporation.

Dedicated line   A telephone line leased to a dedicated customer. These lines are connected to a customer's phone, key telephone system, or PBX.

Digital   A signal encoded as a series of discrete numbers.

Discussion groups   Groups that allow people who have a common interest to electronically share their views and information, do research, or just socialize. Two forms of discussion common on the Internet are public mailings and Usenet newsgroups.

DOS   A single user/single tasking operating system. Only one person can perform one function at a time.

Duplex   A description of the way data bits are transferred in a serial fashion. Duplex (which usually means full-duplex) lets data be transmitted simultaneously in either direction. Half-duplex lets data flow in either direction, but only one way at a time.

Electronic bulletin boards   A system that allows users to communicate by sending and receiving messages and exchanging software and files and provides other services such as games or distribution of data.

Electronic Industries Association   A U.S. organization to which many manufacturers of telecommunications equipment belong.

Electronic mail   Any of a number of ways of sending and receiving messages over an electronic network. Such networks can be private; for users within a company; or public, such as MCI Mail, which is used by subscribers who pay membership fees plus fees for sending messages.

E-mail   See **Electronic mail.**

Ethernet   A widely installed version of a standard used in networking. Ethernet is often used in computer rooms and in terminal-to-mainframe links in offices.

European Conference of Postal and TC Administrations (CEPT)   Founded in 1959, this organization has members from 26 countries. It is a mechanism for coordinating the postal and telecommunications policies of European governments.

Fax   Facsimile. A fax machine transmits graphic matter, such as printing or still pictures, by wire or radio and reproduces it.

FCC   Federal Communications Commission. The board that regulates all interstate and foreign electrical communications systems that originate in the United States: radio, television, facsimile, telegraph, telephone, and cable.

Fiber optics   An increasingly popular cost-efficient communications medium, capable of offering very large transmission capacity. In Japan, Western Europe, and the United States, fiber optics has become competitive with other communications media as the technology of choice in a number of applications.

File server   A disk on a network that acts as the central storage location for applications and data files. Files can be retrieved from the file server, worked on at a workstation, and then saved to the file server. A local area network (LAN) generally has a single, central file server. Some LANs are configured with distributed servers, which are many file servers scattered among workstations.

File transfer protocol (FTP)   A tool that lets users transfer, browse, and download files from one computer on the Internet to another computer connected to the Internet.

Frequency   The number of complete cycles of a periodic activity which occur in a unit time.

Frequently asked questions (FAQ)   Questions and answers many newcomers have about programs on the Internet. FAQs are often compiled and are available to newcomers who sign on to lists.

Gateway   An electronic way to get into the Internet.

Gateway services   The term *gateway services* often is used to describe certain services that may now be provided by the Bell companies under the AT&T consent decree.

Geosynchronous orbit   The path that a geostationary satellite uses. Such a satellite orbits the earth in a path that takes nearly 24 hours for a complete revolution. Consequently, the satellite stays vertically above the same point on the earth's surface—approximately 35,780 kilometers (21,468 miles) over the earth.

Gopher   A menu-driven system which seeks out Internet resources such as archives, libraries, directories, and computers to help users find and retrieve remote files according to subject matter. Unlike FTP, Gopher connects to the host site only to retrieve menus of information and disconnects for browsing.

Hypermedia   A way in which information on the World Wide Web can be connected. Hypermedia allows links to connect words, pictures, sounds, or any type of data file that can be stored on a computer.

Hypertext   A way to read documents on the Internet so that users can read them in any order that makes sense. Readers don't have to read documents from beginning to end. Instead, they can jump around and browse.

Hypertext markup language (HTML)   A protocol that lets a document contain text that can be highlighted and clicked on to jump to another text location, file, or site.

ICA   International Communications Association. Members are primarily companies that are large users of telecommunication services.

Icons   Small picture symbols shown on a computer screen. In the Macintosh user interface, each file stored on a disk is represented by an icon.

Information services   Services often delivered through the public telephone networks to personal computers used as terminals in homes and businesses. They are considered to include consumer videotex, access/retrieval services, messaging services, transactions, personal/environmental management services, computing services, and code and protocol conversion.

International Electrotechnical Commission (IEC)   A nontreaty, voluntary international standards organization formed in 1904. It is concerned with topics like system characteristics, such as voltages, frequencies, and tolerances. In 1987, ISO and IEC formed a joint technical committee on information technology.

International Telegraph and Telephone Committee (CCITT)   A permanent study committee within the 123-year-old International Telecommunications Union, a specialized agency of the United Nations.

Internet   A worldwide network that connects computers at educational, corporate, and government institutions. It brings people of common interests together in electronic communication. Sometimes the Internet is called the Information Superhighway.

Internet relay chat (IRC)   IRC allows people on the Internet to communicate in real time by typing messages.

ISDN   Integrated services digital network. When ISDN is fully implemented, customers should be able to send and receive voice and data communications simultaneously over existing telephone lines.

ISO   International Standards Organization. A group active in developing a standard (open system for interconnection) designed to help hardware and software developers write programs that can easily move between networks.

LAN   Local area network, in which several computers are tied together using telecommunications technology. Large LANs are usually server-based networks, which keep program and data files together on a central hard disk. Most smaller sub-LANs are networks in which each personal computer has its own set of files that only one user can retrieve at any time.

LATA   Local access and transport area. An area served by a single local telephone company. The AT&T consent decree divided the U.S. domestic market into approximately 164 LATAs. Depending upon whether a toll call is interstate (between states) or intrastate (within a state), different regulatory bodies may have jurisdiction over the LATAs.

Least cost routing (LCR)   In least cost routing, outgoing calls are automatically routed over phone lines that provide the lowest cost circuits at the time when the call is placed.

LEC   Local exchange carrier.

Mailing lists   Users who have specific interests can subscribe to mailing lists on a wide variety of subjects by sending e-mail to the computers that maintain them.

Microwaves   Certain radio frequency wavelengths. In past years, analog microwave radio was the primary way long-distance communications were transmitted. However, fiber installations are cutting into the microwave market. Digital microwave radio is being increasingly used for short-distance transmission, especially for point-to-point bypassing of the local telephone company.

MIS   Management information system. Often, a department within an organization.

Modem   A device that translates digital signals from a computer to analog signals that can be transmitted over telephone lines. When the signals reach the receiving computer, another modem translates them back to digital signals. This conversion process lets computers send data to each other. The process is called modulation-demodulation.

Modified Final Judgment (MFJ) A consent decree, effective in 1984, that caused a fundamental restructuring of the U.S. domestic telephone business, set up seven regional Bell holding companies, and placed restrictions on the divested Bell companies' permissible activities.

Multiplex, MUX Using a common channel to make two or more channels. Two common multiplexing techniques are frequency division multiplexing, which uses different carrier frequencies to hold signals, and time division multiplexing, which separates signals by interleaving them.

NetView Proprietary software from IBM that allows network management functions. NetView Plus is similar software that runs on a personal computer, or PC.

Network A way in which signals or data can be shared. Resources are linked together so information can be sent where it is needed. There are many kinds of networks: telephone, video, radio, and satellite communications are some. A local area network (LAN) is one type of network that uses cables, lines, and software to connect various computer devices to send and receive data.

Networking A generic term for software systems that allow taking computers and hooking them together into a multiuser system.

Newsgroups Usenet (also called Net News) is a system of public discussion groups dedicated to a specific topic. There are lists for almost every topic you can think of.

Node In networking, a junction point.

NTIA National Telecommunications and Information Administration, an agency under the U.S. Department of Commerce.

Operating system A system that controls the most basic functions of the computer, including the way in which files are created, stored, accessed, and deleted. Without an operating system, each program would have to control these operations on its own. Operating systems include UNIX, XENIX, DOS, and OS/2. An operating system provides a common interface for programs. Without an operating system, the computer would not work.

OSI Open systems interconnection. A set of standards being developed internationally to make it possible for computer systems to operate compatibly with each other throughout the world.

Packet switching A system that transfers messages in small units (packets) that are individually addressed and sent through a network. In packet switching, the transmission channel is busy only while each packet is being sent. Packet switching is a way in which networks can operate efficiently, since the channel can be used for other traffic after the packet has been sent.

Paging A one-way signal to alert a user to do something. Paging signals, which are very short duration transmissions, are generally tone-only, tone-voice, or alphanumeric messages.

Parity A method used in transferring data that helps to cut down on transmission errors. The total number of 1's or 0's for each character are changed by adding an extra binary signal. Parity is set before transferring the data. Each time a byte is transmitted from one place to another, the byte arriving at the destination is checked automatically for correct parity. If a bit was lost during transmission, parity checking will detect it.

PBX Private branch exchange. A switch, located on a customer's premises, that is used increasingly not only for handling telephone traffic but for handling other kinds of business communications, including all forms of data traffic.

PC Personal computer.

Pixel A picture element or one phosphor or dot on the screen. It is illuminated by an electron gun that sweeps from left to right and top to bottom on the screen. The resolution of a monitor is judged by the number of pixels on the screen.

Protocol Rules for operating a communication system. The way systems talk to each other is defined by protocols, or standards. Protocols work in a hierarchy of various layers. Each layer has different functions. For instance, layer 3 (the network layer) in the OSI model is the common function of the telephone system.

Providers Organizations that provide users with access to the Internet. Many providers sell their services commercially.

Queuing The way in which telephone calls are held in the order of their arrival and sent—in the same order—to an operator or system for handling. For instance, you are put in the queue when you call an airline reservation system or a department store and are put on hold until your turn comes.

RAM   Random-access memory. A computer's internal memory, where a computer temporarily stores data, textfiles, or a worksheet while you work on it.

RBOC   Regional Bell operating company. In 1984, as part of divesture, the AT&T organization was split up. The regional Bell operating companies created as a result of divesture were: Nynex (New York and New England), Bell Atlantic, BellSouth, Southwestern Bell, US West, Pacific Telesis, and Ameritech.

Redundancy   Providing more equipment or signals than are really needed so that if something fails, you have a backup.

Repeater   A device that amplifies a signal or pulse for retransmission.

RS 232 C   An interconnect standard between computer hardware and peripherals, such as a printer.

Satellite   Communications satellites are radio relay stations in orbit above the earth. They are used to provide services between fixed points, between mobile vehicles and fixed points, and to locate the position of vehicles. Most communications satellites are geostationary; some orbit over the poles.

SNA   Systems Network Architecture. A proprietary system developed by IBM.

Star topology   A pattern used in networking to connect terminals and computers. With star topology, all nodes (or terminals) are connected to a central system. The hub, or central point of such a network, directs data passing between workstations and file servers and boosts the electronic signal that carries the data. Networks using twisted pair cable use a hub.

Stop bits   Part of a protocol for transmitting signals by modem. A stop bit lets the receiver of asynchronous transmission come to a halt before accepting another character.

T-1   As used in the United States, the basic 24-channel 1,544-megabits-per-second pulse code modulation system.

Tariff   Rates and rules governing telecommunications services.

Telcos   An industry term for telecommunications companies.

Telecommunications   The wide range of communications and information services that has developed since the invention of the telephone.

Teleconferencing   An electronic, interactive way of communicating at more than two sites by more than two people. Types of teleconferencing include audio, audio graphics, and full motion video, either using analog signals or compressed digital signals.

Telegraph   A method of sending written messages by manual or machine code.

Telex   A worldwide teletypewriter exchange service provided by various companies or government-run agencies.

Telnet, Tymnet   Each is a packet-switching network through which users can connect with the Internet or with certain sites on the Internet.

Token ring   A way in which local area networks are set up to exchange data. A token is passed from station to station in a preset sequence. A station that wants to send a message has to wait till the token gets there before it can transmit data.

Topologies   Various shapes and figures. In networks, the form of the nodes and links. Bus topology connects the nodes in a series on one continuous length of cable. Ring topology connects devices in an unbroken circle. Star topology connects all devices directly to a hub.

Traffic   Telecommunications messages, either sent or received.

Trunk   A circuit between switching centers, switching equipment, or central offices.

Turnkey   An arrangement in which the vendor supplies the entire system, hardware and software. Usually, with a turnkey system, training and service are also supplied by the same vendor.

Twisted pair   Two insulated wires twisted together.

UNIX   A proprietary operating system created by AT&T that is a multi-user/multitasking system.

UNMA   Unified Network Management Architecture, a proprietary system developed by AT&T.

USTA   United States Telephone Association.

VAX   A proprietary mainframe computer from Digital Equipment Corporation (DEC).

Veronica   A search service that allows users to search gopher menus and attached files for words of interest and creates a new gopher menu composed only of the items that appeal to a user's current interest.

Videoconferencing   A type of teleconferencing that simulates a face-to-face meeting. Essential elements of a videoconferencing system include user environments, transmission, and Codec, a device that converts analog signals into digital form for transmission and converts them back again at the destination.

Videotex   A type of electronic information service popular with business users and many home computer users who conduct transactions, send and receive nonentertainment electronic messages, and gain access to a wide range of information services.

Virtual reality   Computer technology that allows users to enter and navigate simulated worlds and environments.

Voice mail   A voice messaging system.

Voice messaging   A technique where spoken messages are recorded for playback when the person to whom the message is being sent becomes available.

Voice response technology   Permits customers to call a special number and receive a series of options, which they access by pressing the telephone keypad on a touch-tone phone. Banks, for instance, can give customers account balances, loan balances, loan rate information, and information on specified transactions by using voice response. Many such systems are available 24 hours a day. Some voice response systems offer bill payment and information services. Voice response units can make up sentences from words—originally spoken by a person and stored by analog or digital technology—or they can use synthetic voice techniques to create sounds.

WANS   Wide area networks. A wide area network is a communications network that serves areas that are geographically separate from each other. (A LAN, or local area network, is also a communications network, but usually within a single building complex.) LANs let a group of computers talk to each other and share data storage devices and printers. Sometimes one LAN communicates with another in a WAN.

WATS   Wide area telephone service. A special telephone company service that lets companies reduce costs of phone calls through certain arrangements for telephone call billing.

Wide area information system (WAIS)   A searching tool that allows users to search many databases on the Internet easily by key word and content. Users can research a variety of topics by typing simple commands.

World Wide Web (WWW)   Internet information service based on hypertext links between various kinds of files: text, graphics, video, and even sound. WWW connects documents to other documents through links (references). The WWW is accessed by programs known as browsers or Webcrawlers.

X.25   A CCITT (International Telegraph and Telephone Consultative Committee) recommendation on protocol for packet switching.

X.400   A CCITT recommendation on the protocol for message handling systems.

RECOMMENDED READING AND RESOURCES

If you are interested in working in telecommunications, keeping up with industry developments is essential. Whether you are already working in the field or considering telecommunications as a possible career, the technology and opportunities are changing so rapidly that you must monitor what is happening. Fortunately, telecommunications is such an important subject that it is well covered by the media. Useful, up-to-date information is as close as your computer—or your nearest library.

CD-ROM SEARCHING

Increasingly, libraries are using CD-ROM technology to help patrons get useful information. Publications like Gale's *Encyclopedia of Associations,* the *Wall Street Journal,* and the Chicago *Tribune* have article collections available on CD-ROM disks that can be computer-searched. Libraries equipped with this technology and terminals that patrons can use often subscribe to services that update the disks periodically.

ON-LINE SEARCHING

If you have access to a computer with a modem and the ability to download to a disk or hard drive, you can find much on-line information quickly. An increasing number of libraries are allowing patrons to dial up information from on-line catalogs—often, after hours and on weekends.

Ask your school or public library how to set your communications software. You may need to alter your settings to emulate a compatible terminal. After your modem has dialed the library's computer and made a satisfactory connection, you can search by subject, author, or title. Once you have located books or publications, you can often find out from your computer screen whether they are on the shelves, can be reserved for you, or can be ordered for you through interlibrary loan.

In metropolitan areas, libraries often have joined a consortium that permits simultaneous searching through a master database. For instance, through Illinet Online (Illinois Library Networks), one such Chicago-based consortium, it is possible—with just a few keystrokes—to look for books owned by more than 800 Illinois libraries.

Illinet also offers services for magazine and journal resources. Many of the periodicals listed below are indexed, and those indexes are on-line. For instance, if you are looking for information on "cellular-telecommunications," you can search *Applied Science and Technology Index,* retrieve 91 citations, sort the listings chronologically with the most recent article listed first or arrange the listings alphabetically by author, and, if you have an e-mail address, send the information automatically to your "mailbox."

Here is a useful tip. Are you looking for "telecommunications" or "telecommunication"? It matters. If you want citations on cellular phones and just type "cellular," you will get biology information unless you specify "cellular-telecommunications." Computer searching looks for exact matches in databases. You will waste time and phone costs for your connection if you don't use the "right" key words.

Consequently, before you start on-line searching, it is helpful to have a printout of the Library of Congress headings for your topic at your fingertips. Then you can search for the appropriate subject. Often subjects aren't obvious. For example, in order to locate magazine and journal citations for the paging industry, you need to ask for "radio paging equipment industry."

If you don't have access to on-line searching, you can usually find material in the indexes described below. Some, like *Readers' Guide to Periodical Literature,* are actually on-line, but access to the information is limited to faculty and students at a particular university.

REFERENCE GUIDES

Readers' Guide

An excellent place to start is with the *Readers' Guide to Periodical Literature.* This well-known reference work lists articles by subject that have appeared in magazines and weeklies. For instance, you will find citations for material under "telecommunication" that have appeared in *Health, Scientific American, Time,* and *Business Week* magazines.

Readers' Guide to Periodical Literature will also refer you to related subjects, such as American Telephone & Telegraph Company, communications satellites, computer networks, Illinois Bell Telephone Company, integrated services digital network, ITT Corporation, MCI Communication Corporation, private networks, telecommuting, teleconferencing, telephone, television broadcasting, value-added networks, and Western Union Corporation.

Under some of these topics, you will find subheads and other references. The entry for "telephone" will suggest that you also see cellular radio, telephone service in hotels and motels, and picture phones. Articles are usually in nontechnical language.

Many of the periodicals mentioned in *Readers' Guide* will be available in your local school or public library.

Business Periodicals Index

Another reference tool similar to *Readers' Guide to Periodical Literature,* but with a slightly different focus, is *Business Periodicals Index.* Articles chosen for listing generally deal with financial, management, or economic issues. You will find citations for topics such as telephone rates, routing systems, equipment and supplies, and telecommunication competition.

Although such familiar sources as *Money* and *Business Week* are indexed in *BPI,* you will also find publications like *Communication News, Datamation, Chain Store Age Executive,* and *Barrons.* Related topics like voice/data integrated systems, facsimile transmission, telecommunications consultants, and private networks are cross-referenced with "telecommunication," so you can easily check your particular topic of interest.

BPI also indexes *Telephony,* a weekly newsmagazine of interest to those working in telecommunications.

Applied Science & Technology Index

A third index you will find valuable, particularly if you are interested in scientific and technical issues, is *Applied Science and Technology Index.* This index offers references and subheads that concentrate on technical developments, such as optical communication systems, signal processing, or telecommunication protocols. A number of European journals are indexed, as are more specialized magazines such as *Water/ Engineering Management* and *Journal of the Acoustical Society of America.* Such publications may not be carried by your local library, and you may have to do some detective work to come up with the articles you would like to read.

SPECIALIZED LIBRARIES

If you live in a large city or major metropolitan area, corporations with headquarters nearby often have their own libraries and a librarian on staff. Your reference librarian at your public library may be able to get special permission for you to visit such a corporate library so that you may take notes and photocopy the articles you want.

You can also call the local headquarters of your regional Bell company or your alternative long-distance carrier. Chances are they will have some of the specialized telephone-related journals or publications.

OTHER INDEXES

Not all materials related to telecommunications are cited in the three indexes mentioned above. Another good source available in many public libraries is *The Magazine Index* and its companion computer-based index *Info-Trac*. By typing the topic "telecommunications" on the *Info-Trac* keyboard, you can get a printout of citations.

WHY READING PERIODICALS IS IMPORTANT

Keeping up with industry developments is very important if you plan to go into telecommunications. But there is another reason for spending time and money to locate periodicals now, before you have made the commitment to telecommunications as a career.

Reading a number of industry-related publications and subscribing to one or two of them are easy ways of helping you decide just how interested you are in telecommunications. By the time you have located and gone through 10 or 20 articles and by the time you have studied 20 issues of *Telephony* and *Telecommunications,* you will know whether you still like learning about the topic. You will gradually pick up the vocabulary, and you will begin to see some of the problems telecommunications professionals are facing and some of the directions in which the industry seems to be going.

The knowledge you gain will help you in coursework, if you decide to enroll in telecommunications classes. Or, if you are ready to look for a job in the industry, the information you have acquired will help you discuss telecommunications issues knowledgeably during interviews.

Besides enabling you to keep up with technical and business-related developments, reading the periodicals helps you in other ways. Almost all of them have a calendar of forthcoming events and telecommunications-related shows, along with a contact address. You can write for additional information on the meetings, workshops, and trade shows. Often, special student discounts are given on registration fees and exhibition attendance. Also, readers who sign up for a show on a form included in a particular magazine may receive a special, lower admission rate.

Some of the magazines also have recruitment ads, either from companies seeking to hire telecommunications professionals directly or from employment agencies that specialize in matching qualified candidates with job openings. Although you may not yet be experienced enough to apply for one of these positions, you certainly can get an idea of necessary qualifications from reading the various ads. You can also contact a recruiter and ask what you need to do to be a viable candidate. Perhaps the companies that are placing the ads have summer jobs, scholarship programs, or grants for which you can apply as a telecommunications student. Usually, they are glad to learn of young people interested in the field.

From reading the periodicals and checking the ads in them, you may often learn which companies advertise telecommunications jobs on-line. You can find out more about on-line searching on the Internet in Chapter 12.

MAGAZINES

Applied Optics
  Optical Society of America
  1816 Jefferson Pl. NW
  Washington, DC 20036

Applied Physics Letters
  American Institute of Physics
  335 E. 45th St.
  New York, NY 10017

AT&T Technical Journal
  AT&T Bell Laboratories
  600 Mountain Ave.
  Room 3C443, Box 636
  Murray Hill, NJ 07974

Audiotex News
  2362 Hempstead Turnpike
  East Meadow, NY 11554

Business Communications Review
  BCR Enterprises
  950 York Rd.
  Hinsdale, IL 60521

Business Week
  McGraw Hill Inc.
  1221 Avenue of the Americas
  New York, NY 10020

CB
  Dempa Publications, Inc.
  275 Madison Ave., 32d Floor
  New York, NY 10016

Cellular Business
  Intertec Publishing Corp.
  9800 Metcalf
  Overland Park, KS 66212

Cellular Phone
  Dempa Publications, Inc.
  275 Madison Ave., 32d Floor
  New York, NY 10016

Common Carrier Week
  Warren Publishing
  2115 Ward Ct. NW
  Washington, DC 20037

Communications Daily
  Warren Publishing
  2115 Ward Ct. NW
  Washington, DC 20037

Communications News
  Nelson Publishing Co.
  2504 N. Tamiami Trail
  Nokomis, FL 34275

Computer World
  375 Cochuituate Rd.
  Farmingham, MA 01701

Computerworld Newspaper
  CW Communications Inc.
  529 14th St. NW #650
  Washington, DC 20045

Cordless Phone
  Dempa Publications, Inc.
  275 Madison Ave., 32d Floor
  New York, NY 10016

Data Communications Product Directory
  Architecture Technology
  Box 24344
  Minneapolis, MN 55424

Datamation
  Cahners Publishing
  275 Washington St.
  Newton, MA 02158-1630

Direct Dial
  United Advertising Publications
  15400 Knoll Trail, Ste. 500
  Dallas, TX 75248

DM News
  19 W. 21st Street
  New York, NY 10010

European Mobile Communications
  Holt Cottage
  Kingston Hill
  Kingston-upon-Thames
  KT2 7JH
  England

European Telecommunications
    Probe Research Inc.
    Three Wing Drive, Ste. 240
    Cedar Knolls, NJ 07927

Fiberoptic Product News
    Gordon Publications Inc.
    301 Gibralter Dr.
    Box 650
    Morris Plains, NJ 07950-0650

Forbes ASAP
    (supplement on the Information Age)
    60 Fifth Ave.
    New York, NY 10011-8882

Fortune
    Time Warner Inc.
    Time Life Bldg.
    Rockefeller Center
    New York, NY 10019

411 Newsletter
    United Communications Group
    11300 Rockville Pike, Ste. 1100
    Rockville, MD 20852

Ham Radio
    Dempa Publications, Inc.
    275 Madison Ave., 32d Floor
    New York, NY 10016

Home Office Computing
    411 Lafayette St.
    New York, NY 10003

Information Industry Bulletin
    Digital Information Group
    51 Bank St.
    Stamford, CT 06902

Information Week
    600 Community Dr.
    Manhasset, NY 11030

InfoWorld
    4038 128th Ave. SE
    #312 B
    Bellevue, WA 98006

Internet World
    PO Box 713
    Mt. Morris, IL 61054

ISDN Report
    Probe Research Inc.
    Three Wing Dr., Ste. 240
    Cedar Knolls, NJ 07927

JEE (Journal of Electronic Engineering)
    Dempa Publications Inc.
    275 Madison Ave., 32d Floor
    New York, NY 10016

LAN Magazine
    600 Harrison St.
    San Francisco, CA 94107

L A N Times
    McGraw-Hill, Inc.
    1221 Avenue of the Americas
    New York, NY 10020

Land Mobile Radio News
    Phillips Publishing
    7811 Montrose Rd.
    Potomac, MD 20854

Link-Up
    Learned Information Inc.
    143 Old Marlton Pike
    Medford, NJ 08055

Mac Week
    Ziff Davis Publishing Co.
    301 Howard St., 15th Floor
    San Francisco, CA 94105

Mobile Office
    470 Park Ave. S., 14th Floor S.
    New York, NY 10016

NARTE News
   National Association of Radio & Telecommunications Engineers
   PO Box 678
   Medway, MA 02053

NetGuide
   600 Community Dr.
   Manhasset, NY 11030

Network World
   161 Worcester Rd., 5th Floor
   Framingham, MA 01701

OEP (Office Equipment & Products)
   Dempa Publications Inc.
   275 Madison Ave., 32d Floor
   New York, NY 10016

Office Systems Magazine
   PO Box 150
   Georgetown, CT 06829-0150

Online Access
   Chicago Fine Print Inc.
   900 N. Franklin St., Suite 310
   Chicago, IL 60610

Optical Engineering
   International Society for Optical Engineering (SPIE)
   PO Box 10
   Bellingham, WA 98227

Optics Letters
   Optical Society of America
   2010 Massachusetts Ave. NW
   Washington, DC 20036

PC Computing
   PO Box 58229
   Boulder, CO 80322-8229

Phone Plus
   Taurus Publishing Company
   PO Box C-5400
   Scottsdale, AZ 85261

Satellite TV Finance
   Financial Times Business Information
   126 Jermyn St.
   London SW 1Y4 UJ England

Satellite Week Newsletter
   Warren Publishing
   2115 Ward Ct. NW
   Washington, DC 20037

TE&M's Telecom Asia
   7500 Old Oak Blvd.
   Cleveland, OH 44130

Telecom Gear
   United Advertising Publications, Inc.
   15400 Knoll Trail, Ste. 500
   Dallas, TX 75248

Telecommunications Alert
   United Communications Group
   11300 Rockville Pike, Ste. 1100
   Rockville, MD 20852

Telecommunications Reports
   1333 H St. NW
   Washington, DC 20005

Teleconnect
   12 W. 21st St.
   New York, NY 10166

Telephone Angles
   United Communications Group
   1300 Rockville Pike, Ste. 1100
   Rockville, MD 20852

Telephone Engineer and Management
   Advanstar Communications, Inc.
   7500 Old Oak Blvd.
   Cleveland, OH 44130

Telephony
   Intertec Publishing Corp.
   9800 Metcalk
   Overland Park, KS 66202

Voice Processing Inc.
  Probe Research Inc.
  Three Wing Dr., Ste. 240
  Cedar Knolls, NJ 07927

BOOKS

The list below includes some of the most interesting and useful books if you are considering a career in telecommunications. Checking the *Subject Guide to Books in Print,* available in print at the reference desk in your local library or on CD-ROM, will give you additional titles. Look not only at telecommunications books but also at those listed under the cross-referenced headings. Also check the Library of Congress subject headings under "telecommunications" before you do on-line searching with the help of your reference librarian. You may need to look at additional entries that are cross-referenced.

Acampora, Anthony S. *An Introduction to Broadband Networks: LANs, MANs, ATM, ISDN, and Optical Networks for Integrated Multimedia Telecommunications* (New York: Plenum Press, 1994).

Alspach, Ted. *Internet E-Mail Quick Tour* (Chapel Hill, N.C.: Ventana Press, 1995).

Basye, Anne. *Opportunities in Telemarketing Careers* (Lincolnwood, Ill.: VGM Career Horizons, 1994).

Berek, Britton and Marilyn Canna. *Telemedicine on the Move: Health Care Heads Down the Information Superhighway* (Chicago, Ill.: American Hospital Association, 1994).

Bernard, Josef. *The Cellular Connection: A Guide to Cellular Telephones* (Mendocino, Calif.: Quantum Publishing, 1994).

Bigelow, Stephen J. *Telephone Repair Illustrated* (Blue Ridge Summit, Penn.: TAB Books, 1993).

———. *Understanding Telephone Electronics,* 3d ed. (Carmel, Ind.: SAMS, 1991).

Black, Uyless D. *Emerging Communications Technologies* (Englewood Cliffs, N.J.: PTR Prentice Hall, 1994).

Blum, Daniel J. and David M. Litwack. *The E-mail Frontier: Emerging Markets and Evolving Technologies* (Reading, Mass.: Addison-Wesley Publishing Co., 1995).

Browne, Steve. *The Internet via Mosaic and World-Wide Web* (Emeryville, Calif.: Ziff-Davis Press, 1994).

Bush, Stanley E. *Private Branch Exchange Systems and Applications* (New York, N.Y.: McGraw-Hill, 1994).

Campbell, Becky and Micky Applebaum. *Don't Panic: It's Only Netware* (Indianapolis, Ind.: New Riders Publishing, 1993).

Comer, Douglas E. *The Internet Book: Everything You Need to Know about Computer Networking and How the Internet Works* (Englewood Cliffs, N.J.: Prentice Hall, 1993).

Communications Publishing Service. *Cellular Travel Guide* (Mercer Island, Wash.: Communications Publishing, 1993).

Corrigan, Patrick. *LAN Disaster Prevention and Recovery* (Englewood Cliffs, N.J.: PTR Prentice Hall, 1994).

Cronin, Mary J. *Doing Business on the Internet: How the Electronic Highway Is Transforming American Companies* (New York, N.Y.: Van Nostrand Reinhold, 1994).

Daniels, N. Caroline. *Information Technology: The Management Challenge* (Reading, Mass.: Addison-Wesley, 1994).

Dempsey, Jack L. *Telecom Basics* (Chicago, Ill.: Telephony, 1988).

Deniz, Dervish. *ISDN and Its Application to LAN Inter-connection* (New York, N.Y.: McGraw-Hill, 1994).

Derfler, Frank. *PC Magazine Guide to Linking LANS* (Emeryville, Calif.: Ziff-Davis Press, 1992).

Dougherty, Dale. *The Mosaic Handbook for Macintosh,* 2d ed. (Sebastopol, Calif.: O'Reilly and Associates, 1995).

Dougherty, Dale and Richard Koman. *The Mosaic Handbook: for Microsoft Windows,* 2d ed. (Sebastopol, Calif.: O'Reilly and Associates, 1995).

Dvorak, John and Nick Annis. *Dvorak's Guide to Desktop Telecommunications* (Berkeley, Calif.: Osborne McGraw-Hill, 1990).

Eager, William. *The Information Payoff: The Manager's Concise Guide to Making PC Communications Work* (Englewood Cliffs, N.J.: PTR Prentice Hall, 1995).

Eddings, Joshua. *How the Internet Works* (Emeryville, Calif.: Ziff-Davis Press, 1994).

Ellsworth, Jill. *Education on the Internet* (Indianapolis, Ind.: SAMS Publishing, 1994).

Engst, Adam C. *Internet Starter Kit for Macintosh* (Indianapolis, Ind.: Hayden Books, 1993).

Engst, Adam C., Corwin S. Low, and Michael A. Simon. *Internet Starter Kit for Windows* (New York, N.Y.: Windcrest/McGraw-Hill, 1994).

Estrada, Susan. *Connecting to the Internet* (Sebastopol, Calif.: O'Reilly and Associates, 1993).

Fisher, Sharon. *Riding the Internet Highway* (Carmel, Ind.: New Riders Publishing, 1993).

Fraase, Michael. *The Mac Internet Tour Guide: Cruising the Internet the Easy Way* (Chapel Hill, N.C.: Ventana Press, 1993).

Freestone, Julie and Janet Brusse. *Telemarketing Basics: A Practical Guide for Professional Results* (Los Alton, Calif.: Crisp Publications, Inc., 1989).

Frenzel, Louis E. *Communication Electronics* (New York, N.Y.: Glencoe, 1995).

Frey, Donnalyn. *Directory of E-Mail Addresses and Networks,* 4th ed. (Sebastopol, Calif.: O'Reilly and Associates, 1994).

Gardner, James. *A DOS User's Guide to the Internet: E-Mail, Netnews, and File Transfer with UUCP* (Englewood Cliffs, N.J.: PTR Prentice Hall, 1994).

Gibbs, Mark. *Absolute Beginner's Guide to Networking* (Indianapolis, Ind.: SAMS Publishing, 1995).

Gilster, Paul. *Finding It on the Internet: The Essential Guide to Archie, Veronica, Gopher, WAIS, WWW (including Mosaic), and Other Search and Browsing Tools* (New York, N.Y.: John Wiley & Sons, Inc., 1994).

———. *The Internet Navigator* (New York, N.Y.: John Wiley & Sons, Inc., 1993).

Glossbrenner, A. *Internet 101: College Student's Guide* (New York, N.Y.: Windcrest-McGraw-Hill, 1995).

Godin, Seth. *eMarketing* (New York, N.Y.: Berkeley Publishing Group, 1995).

Gross, Lynne S. *Telecommunications: An Introduction to Electric Media,* 5th ed. (Dubuque, Iowa: Wm. C. Brown & Benchmark, 1995).

Hahn, Harley and Rick Stout. *The Internet Yellow Pages* (Berkeley, Calif.: Osborne McGraw-Hill, 1994).

Hanson, Janice. *Connections: Technologies of Communication* (New York, N.Y.: HarperCollins College Publishers, 1994).

Held, Gilbert. *The Complete Cyberspace Reference and Directory* (New York, N.Y.: Van Nostrand Reinhold, 1994).

————. *The Complete Modem Reference: The Technician's Guide to Installation, Testing, and Trouble-free Telecommunications,* 2d ed. (New York, N.Y.: John Wiley & Sons, 1994).

————. *Local Area Network Performance: Issues and Answers* (New York, N.Y.: John Wiley & Sons, 1994).

Heldman, Robert K. *Information Telecommunications: Networks, Products, & Services* (New York, N.Y.: McGraw-Hill, 1994).

Hoffman, Paul E. *Internet Instant Reference* (Alameda, Calif.: Sybex, 1994).

Horowitz, Robert Britt. *The Irony of Regulatory Reform: The Deregulation of American Telecommunications* (New York, N.Y.: Oxford University Press, 1989).

Kent, Peter. *The Complete Idiot's Guide to the Internet* (Indianapolis, Ind.: Alpha Books, 1994).

————. *10 Minute Guide to the Internet* (Indianapolis, Ind.: Alpha Books, 1994).

Khan, Ahmed S. *The Telecommunications Fact Book and Illustrated Dictionary* (Albany, N.Y.: Delmar Publishers, 1992).

Krol, Ed. *The Whole Internet Users Guide and Catalog* (Sebastopol, Calif.: O'Reilly and Associates, 1994).

Kugelmass, Joel. *Telecommuting: A Manager's Guide to Flexible Work Arrangements* (New York, N.Y.: Lexington Books, 1995).

Lambert, Steve and Walt Howe. *Internet Basics: Your Online Access to the Global Electronic Superhighway* (New York, N.Y.: Random House Electronic Publishing, 1993).

LaQuey, Tracy L. with Jeanne C. Ryer. *Internet Companion: A Beginner's Guide to Global Networking* (Reading, Mass.: Addison-Wesley, 1993).

Lawley, Elizabeth and Craig Summerhill. *Internet Primer for Information Professionals* (Westport, Conn.: Mecklermedia, 1993).

Levine, John R. and Carol Baroudi. *Internet for Dummies* (San Mateo, Calif.: IDG Books, 1993).

Levine, Toby Kleban. *Going the Distance: A Handbook for Developing Degree Programs Using Television Courses and Telecommunications Technologies* (Alexandria, Va.: Annenberg/CPB Project, PBS Adult Learning Service, 1992).

Lowe, Doug. *Networking for Dummies* (San Mateo, Calif.: IDG Books, 1994).

Madron, Thomas William. *Local Area Networks: New Technologies, Emerging Standards,* 3d ed. (New York, N.Y.: Wiley, 1994).

Meckler, Alan. *On Internet 95: An International Guide to Resources on the Internet* (Westport, Conn.: Mecklermedia, 1995).

Midwinter, John E., ed. *Photonics in Switching* (Boston, Mass.: Academic Press, 1993).

Minoli, Daniel. *Analyzing Outsourcing: Reengineering Information and Communication Systems* (New York, N.Y.: McGraw-Hill, 1995).

Morgan, Eric Lesse. *WAIS and Gopher Servers: A Guide for Internet End-Users* (Westport, Conn.: Mecklermedia, 1994).

Nilles, Jack M. *Making Telecommuting Happen: A Guide for Telemanagers and Telecommuters* (New York, N.Y.: Van Nostrand Reinhold, 1994).

Notess, Greg R. *Internet Access Providers* (Westport, Conn.: Mecklermedia, 1994).

Peters, Meike and Joyce Whiting. *Novell's Quick Access Guide to NetWare 3.12 Networks* (San Jose, Calif.: Novell Press, 1993).

Pierce, John Robinson. *Signals: The Science of Telecommunications* (New York, N.Y.: Scientific American Library, 1990).

Quarterman, John. *The E-mail Companion: Communicating Effectively via the Internet and Other Global Networks* (Reading, Mass.: Addison-Wesley Publishing Co., 1994).

Rees, David W. E. *Satellite Communications: The First Quarter Century of Service* (New York, N.Y.: John Wiley & Sons Publishing, 1989).

Reid, Alastair. *Teleworking: A Guide to Good Practice* (Cambridge, Mass.: NCC Blackwell, 1994).

Reoffey, Lynne. *Modem USA* (Washington, DC.: Allium Press, 1994).

Resnick, Rosalind and Dave Taylor. *The Internet Business Guide: Riding the Information Superhighway to Profit* (Indianapolis, Ind.: SAMS, 1994).

Rulten, Peter. *Net Guide: Your Map to the Services, Information and Entertainment on the Electronic Highway* (New York, N.Y.: Random House, 1994).

Sachs, David. *Hands-on Internet: A Beginning Guide for PC Users* (Englewood Cliffs, N.J.: PTR Prentice Hall, 1994).

Schepp, Brad. *The Telecommuter's Handbook: How to Work for a Salary— without ever Leaving the House* (New York, N.Y.: Pharos Books, 1990).

Schneiderman, Ron. *Wireless Personal Communications: The Future of Talk* (New York, N.Y.: IIEE Press, 1994).

Seybold, Andrew M. *Using Wireless Communications in Business* (New York, N.Y.: Van Nostrand Reinhold, 1994).

Shafiroff, Martin D. and Robert L. Shook. *Successful Telephone Selling in the '90s* (New York, N.Y.: Perennial Library, 1990).

Shannon, Larry R. and Janet Shannon. *Welcome to Home-Based Business Computing* (New York, N.Y.: MIS Press, 1995).

Shirkey, Clay. *The Internet by E-Mail* (Emeryville, Calif.: Ziff-Davis Press, 1994).

Skurzynski, Gloria. *Telecommunications in Your High-Tech World* (New York, N.Y.: Maxwell Macmillan International, 1993).

Smith, Carter. *America's Fastest Growing Employers: The Complete Guide to Finding Jobs with Over 700 of America's Hottest Companies* (Holbrook, Mass.: Bob Adams, Inc. 1992).

Snider, Jim and Terra Ziporyn. *Future Shop: How New Technologies Will Change the Way We Shop and What We Buy* (New York, N.Y.: St. Martin's Press, 1992).

Stallings, Will. *The Business Guide to Local Area Networks* (Carmel, Ind.: Howard W. Sams, 1990).

Stone, Alan. *Wrong Number: The Breakup of AT&T* (New York, N.Y.: Basic Books, 1989).

Tedesco, Eleanor. *Telecommunications for Business* (Boston, Mass.: PWS-Kent Publishing Co., 1990).

Tittel, Ed and Margaret Robbins. *E-Mail Essentials* (Boston, Mass.: AP Professional, 1994).

Tuttlebee, Wally H. W., ed. *Cordless Telecommunications in Europe: The Evolution of Personal Communications* (London, England: Springer-Verlag, 1990).

Vaskevitch, David. *Client/Server Strategies: A Survival Guide for Corporate Engineers* (San Mateo, Calif.: IDG Books, 1993).

White, Gordon. *Mobile Radio Technology* (Boston, Mass.: B-H Newnes, 1994).

Yoffie, David B. *Strategic Management in Information Technology* (Englewood Cliffs, N.J.: Prentice Hall, 1994).

Young, Paul H. *Electronic Communication Techniques* (New York, N.Y.: Merrill, 1994).

ASSOCIATIONS

Because telecommunications is changing so rapidly, one of the best ways to stay current in knowledge and technology is to be aware of what associations are doing. Some, listed below, are manufacturers' organizations. Others may have discounts for student memberships. Still others offer publications for sale as well as conferences, seminars, and workshops for members. Sometimes students can attend such activities at reduced fees.

Additional associations are listed in Chapter 13.

American Telemarketing Association
444 N. Larchmont Blvd., Ste. 200
Los Angeles, CA 90004

Association of Communications Technicians
1501 Duke St.
Alexandria, VA 22314

Association of Telemessaging Services International
1150 S. Washington St., Ste. 150
Alexandria, VA 22314

Electronic Industries Association
2001 Pennsylvania Ave. NW
Washington, DC 20006-1813

Industrial Telecommunications Association
1110 N. Glebe Rd., Ste. 500
Arlington, VA 22201

Interactive Services Association
8403 Colesville Rd., Ste. 865
Silver Spring, MD 20910

International Communications Association (ICA)
    12750 Merit Dr., Ste. 710, LB-89
    Dallas, TX 75251-1240

International Facsimile Association
    4019 Lakeview Dr.
    Lake Havasu City, AZ 86403

International Teleconferencing Association
    1150 Connecticut Ave. NW, Ste. 1050
    Washington, DC 20036

International University Consortium
    University of Maryland
    University College
    College Park, MD 20742

InterUniversity Communications Council
    1112 16th St. NW, Ste. 600
    Washington, DC 20036

National Association of Business and Educational Radio
    1501 Duke St.
    Alexandria, VA 22314

National Association of Cellular Agents
    1716 Woodhead St.
    Houston, TX 77019

National Communications Association
    16 E. 34th St., 15th Floor
    New York, NY 10016

Society of Telecommunications Consultants
    23123 S. State Rd. 7, Ste. 220
    Boca Raton, FL 33428

Telecommunications Industry Association
    2001 Pennsylvania Ave. NW, Ste. 800
    Washington, DC 20006

Telocator, The Personal Communications Industry Association
    1019 19th St. NW, Ste. 1100
    Washington, DC 20036

Wireless Cable Association
    1155 Connecticut Ave. NW, Ste. 700
    Washington, DC 20036